Workforce & Business Development Center

A Disruptive Innovation for Sustainable Economic Recovery

WORKFORCE & BUSINESS DEVELOPMENT CENTER

A DISRUPTIVE INNOVATION FOR SUSTAINABLE ECONOMIC RECOVERY

EII
PUBLICATIONS

Robert K. Wysocki

Workforce & Business Development Center
A Disruptive Innovation for Sustainable Economic Recovery
All Rights Reserved.
Copyright © 2010 Robert K. Wysocki
v2.0

EII Publications
www.eiicorp.com

ISBN: 978-1-933788-36-4

PRINTED IN THE UNITED STATES OF AMERICA

Introduction

The American Worker has been cut off and is adrift in the mainstream economy. In the spring of 2010 over 17% were unemployed, underemployed, trying to enter the job market for the first time or to re-enter after several years of separation. To that 17% I would add many others that do not have the needed technical skills, have lost hope and are no longer actively searching and the actual figure rises to well above 20%. The American Workers' dreams of starting their own businesses are clouded in confusion and uncertainty. Dwarfing all of this is their concern for their long-term survival as a worker or entrepreneur in this volatile and unpredictable economy. There is a bewildering array of career directions to choose from but which ones will still be viable in two, three, or five years? Even though the economy may be rebounding there is a continuing concern for choosing a career path that offers some protection against another weak job market following upon the heels of a repeat of the economic disaster. The American Worker needs help and advice.

At the same time we know that the American Worker is creative, tenacious, and rises to challenges. The potential of the worker is clear but their separation from the economy is a barrier that must be removed before any sustainable progress can be made. Unfortunately the training and education communities that should be there for them are only passively involved and are not prepared to take on the challenging array of problems that are more complex than any they have ever faced in their professional lives. The training and education community needs guidance and help in designing and implementing programs to prepare the American Worker for job entry.

The American Entrepreneur has also been cut off and is adrift in the mainstream economy. Their biggest issue is whether or not their business idea has any staying power. Can their idea make money for them and if so, what is the best way to make it happen? If they think they are putting themselves in harm's way and at risk of their idea being replicated by an off-shore company, what should they do? What can they do? They need a positioning strategy. Is there a strategy that will afford them the staying power for their business that they so desperately must have? Can they leverage technology to insulate their business idea from the entry of new competitors? They need help!

Business incubation centers (BIC) are located in every state but the services they offer are more of the passive operational type of support than they are of the proactive strategic support that is needed. The Small Business Administration (SBA) was formed in 1953 and there are now over 1100 local Small Business Development Centers (SBDC). They can be found in most colleges and universities. Unfortunately their support is more reactive than proactive. They point the entrepreneur in directions with all kinds of information that may answer their questions but by and large the entrepreneur is on their own. SBDCs offer consulting services using student teams under the guidance of faculty members but the services tend to be more cursory than meaningful. The entrepreneur needs a different kind of help than is available! They need a support structure that creates a true partnership with them that is focused on the success of their business venture. They need an objective evaluation of their idea and strategic advice on how or even if they should proceed.

I would be remiss if I didn't recognize the interdependence between the American Worker and the American Entrepreneur. Many times they are the same person and have decided to launch their own business to get back into or enter for the first time the mainstream of the economy. Their needs range from the very fundamental understanding of business formation and development to a complete dependence on outside assistance. These needs dwarf the separate needs of the American Worker and the American Entrepreneur. Oftentimes they will be novices with respect to entrepreneurship. They need help and there isn't any organized one-stop service they can turn to.

It's time to introduce a disruptive innovation in workforce and business development and that is the purpose of this book.

Background

I'm not a credentialed educational researcher but I have designed, developed and taught credit and noncredit courses to thousands of undergraduate and graduate students and trained thousands of working professionals all over the world for the past 40 years. My career spans 20 years in higher education and 20 years managing my own training and consulting practice. I have worked successfully on both sides of the aisle. Over the years I have continuously put a high priority on delivering products and processes for the vocational, career planning and business development needs of individuals from high school through graduate school and beyond and to my business clients from all over the world. I've been told I'm pretty good at it too. So I think I've earned the right to pass on to the experts some observations I've made and conclusions I've drawn about what constitutes an effective workforce and business development center (WBDC). A word of caution is in order though. My WBDC is not to be confused with your father's workforce development program (WDP). They are similar in purpose but completely different in structure. Most WDPs offer instructor-led programs and not much else. The WBDC, on the other hand, offers project-driven curriculum delivered to teams as but one of its many services. It is unique and offers an innovative approach to developing workers, entrepreneurs and business owners for all ages and times. The weakened state of the global economy and high unemployment rates makes this a timely topic.

Let me put my stake in the ground right now by saying that based on my criteria and extensive research I haven't found any effective long-term WBDC Models operating, at least not in the U.S. I'm basing that statement on a holistic view of the problem environment that includes workforce training and development, business formation, and business development and the supporting infrastructure to effectively help both groups. This holistic view of the problem embraces students from high school through graduate school and beyond as well as businesses and business processes from formation through maturity. I will show you how workers, entrepreneurs and business owners are inextricably linked to the problem and also essential contributors to its solution. There is an untapped potential in the learning-business-student relationships and taking advantage of that potential is my focus in this book.

Many educational researchers share my opinion that the instructor-driven models used for WDPs in the post-secondary markets are broken. The breakage is so severe that these programs can't be fixed by just making incremental changes. These programs can only be fixed by a wholesale replacement of instructor-driven models by team-driven models and the integration of project/problem-centric curriculum where appropriate. These new models are truly revolutionary in that the student teams drive the learning experience not the instructor. The role of the instructor becomes as much a facilitator as a teacher.

Current Workforce Development Programs can only be fixed by a complete replacement of instructor-centric models with team-centric models and the integration of project/problem-based curricula.

That is the direction I believe we need to take in the development of effective WBDC Models and the direction that I am championing in this groundbreaking book. I have designed and am documenting for the first time in this book a project-based team-centric WBDC Model that I know from plain old common sense reasoning and personal experience will help the American Worker get back on their feet and stay on their feet. The term workforce development has a lot of baggage mostly stemming from its dependence on instructor-based training and development. The WBDC Model that I am proposing should not be confused with those models. It is much more than that. While the model is conceptual it is also practical and is just beginning to be implemented in the U.S. and internationally. The book includes an implementation plan for a WBDC Model that you will be able to customize to exactly fit your environment and market position.

The WBDC Model is not for the faint of heart. I have firsthand experience in several different academic settings and I know that for many of you this will be a big step outside of your comfort zone. For others it will be a welcome change. For both audiences it will require taking a very different look at training and education. That is a big step outside of your comfort zone as are most disruptive innovations. The only comfort I can offer you is that it is the right thing to do and, if you follow my lead, it is guaranteed to work. In the pages that follow I will describe the WBDC Model and give you a practical plan for customizing and implementing it.

A Fundamental Reality of the Business World

Most contemporary organizations operate in a project-based cross-functional team-driven mode. The stove pipe models with their "throw it over the wall" mentality have been obsolete for decades. In their place are non-hierarchical structures that operate within a team structure on projects and business processes. Virtually everything in a company can be reduced to either a project or a process and the WBDC addresses both with comprehensive products and services. Furthermore, the projects and processes assigned to teams are no longer simple. They are complex and require multiple disciplines and hence provide the rationale for a team-driven effort.

Doesn't it follow then that the professional development programs for the company's workers would be more meaningful and motivating if they were based on a team-driven approach to education, training and development? It sounds like a no-brainer to me. Corporate Universities are positioned to take advantage of such programs. Unfortunately to have the education community break with the obsolete traditional lecture format and adopt the new format is no easy task. There is great comfort in continuing the same old practices despite the fact that they are not meeting contemporary needs. I hope to break that chain of thinking with my WBDC Model.

Working as the member of a team presents the individual with challenges they do not encounter when working as an individual contributor. Learning how to accommodate differences in work habits, handling conflict situations, problem solving and decision making are but a few of those challenges. Most important of all is the synergistic effect that follows from having several minds focus on the project or problem rather than just one mind. Team members will learn from each other and create a powerful result. Projects will be completed and problems solved that otherwise would not have been had it not been for the power of the team. It makes sense to me that to be most effective a WBDC Model should reflect the work environment. My WBDC Model mimics this work environment as much as any learning and training experience can. As you read and learn about the WBDC Model I hope you will agree with me.

So I ask you to keep an open mind as you consider the possibilities of the WBDC Model that I am advocating. Decide for yourself whether or not I am on the right track. In my mind what I am advocating is just good old common sense and is based on 35 years of experimentation, experience and observation. The WBDC wasn't born fully matured like Venus on the half shell. Instead it is the product of several iterations. My WBDC Model is an innovative disruption. Planning and implementation will require significant change and effort because it goes to the very core of long-standing educational mores. Subject matter experts can easily become entrenched in their viewpoint and resistant to change. Change often takes people outside of their comfort zone and so they are naturally resistant.

We have reached the point in the history and performance of our educational delivery systems where the pain of our failures far outweighs the pain of change and it's time to move ahead courageously.

I believe that a well-thought out implementation plan can be developed. I know how to do that and leave that as the topic of discussion in Chapter 7.

Even though I've given the WBDC Model considerable thought it is still a work in process. Some things can only be learned through practice and the ideal WBDC Model is one of those things. It will be different for every organization that implements it. As is true of any disruptive innovation much remains to be done but at least I can offer a place to start by documenting my cumulative experiences dating back to the early 1970s with the different parts of the Model. Whatever the future holds for this Model can only be guessed at this time but it will be continuously updated in future editions of this book.

THE FORMATION OF THE WBDC MODEL

My professional career spans 40 years and during that time I have always focused on helping the individual plan and pursue programs for their career and professional development. I have designed and developed a number of processes and programs to accomplish that purpose. In this section I recount those efforts and how they eventually led to the formation of the WBDC Model. What follows then are snippets from my 40 years experiences in business and academe that framed my thinking about the WBDC Model.

1970 IT Staff Development

My first position as a people manager was as Director of Computing at a state supported university in Louisiana. The department was small (about 30 professionals) and all were civil service employees. Annual turnover was about 30% and it seriously compromised my ability to consistently offer the service and support expected of my user community. It seemed that I would no longer train a staff member than I would lose them. Low salaries and lack of career growth opportunities were the two major reasons for the high turnover. I couldn't do anything about low salaries but I could do something to increase the value of their work experiences. If the perceived value of the work experience exceeded the value of the salary, I could reduce annual turnover.

Early one morning Ed, the newly hired third shift operator, came into my office. He was recently discharged from the Army. Passed the civil service computer operator's test and with his veteran's preference was first on the list of qualified applicants. By law I had to offer him the position. This was his first real job since leaving the service. He was now three months on the job and he told me that he often sat alongside the programmers and watched them debug their programs. He was interested in being a programmer and asked me what he could do. I asked him if he knew any programming languages. He didn't but he wanted to learn. I gave him a simple report to write, told him to use a popular report generator program that we were using and gave him the necessary manuals. A few weeks later he came back and showed me the report he had produced. "Is this what you wanted?" he asked. "Exactly", I said. Several such successfully completed assignments later I was able to promote him to an assistant programmer position. He really blossomed!

I learned a valuable lesson from that experience. Most people respond to a challenge as long as they have the means to take on that challenge and master it and once they have mastered it know that there is another one to take on. I used that model and was able to promote a data entry clerk to first shift supervisor and a programmer to systems analyst. Within three years annual turnover in the department was down to 7%! Furthermore, there was an alignment between the career objectives of the employee and the service delivery objectives of the department. That is essential to creating a win-win situation!

Most people respond to a challenge as long as they have the means to take on that challenge and master it and once they have mastered it know that there is another one to take on.

Technical professionals responded positively to challenges and if I could align their assignments with their career and professional development interests, I might be able to create a win-win situation. I win by reducing turnover and they win by progressing along a path that matches their career interests. So here is the professional development plan (PDP) process that I initiated:

- Step #1: Establish the PDP
- Step #2: Establish Annual Plan
- Step #3: Measure progress against the Annual Plan
- Step #4: Update the PDP
- Step #5: Return to Step #2

The lesson for me was that motivation and opportunity are keys to career progression and professional development. See Chapter 6 for a complete discussion of the PDP and its use in professional development.

1974 Washington HS

I was privileged to be part of the advisory committee for one of the country's earliest magnet school programs at Washington High School in Milwaukee, Wisconsin. This magnet public school was located in one of the most depressed neighborhoods in Milwaukee and would be a special challenge. The Washington High School program specialized in business and information systems. In addition to coursework the program featured captive businesses run by students. I recall a bakery, a graphic design business and a copy center business. The businesses had simple information requirements and the business information systems students designed, developed and implemented the computer systems support for the businesses. The interesting feature was that these three captive businesses drove the need for systems – a simple project-driven curriculum model. This approach created a high level of motivation of the students in the program. I recall talking to one of the students at the close of the first year in the program. I asked "what was the most valuable part of the year for you?" He enthusiastically replied "I found out that there really are useful applications of algebra!" They were amazed that they could actually find a reason to learn algebra and because they needed it they were highly motivated to learn it. This is the same principal that motivated the operator who wanted to become a programmer in the previous section.

The lesson for me was that learning can be motivated by knowing how that learning will be used to meet a specific and important need.

1980 Business/Academe Partnerships

Business/Academe partnerships have long been discussed as a way to enhance the learning experiences and motivation of students. But like the weather there was a lot of talk but no one seemed to be able to do anything about it. In this example Washington High School was the client and the Management Computer Systems students at the University of Wisconsin-Whitewater were the consultants. At that time I was the Associate Dean in the College of Business and acted as account manager for the engagement. The students earned a small salary for working on the project. The project was to develop an alumni tracking system for the Washington High School magnet program. The project spanned two academic years and involved about 15 university students.

The lesson for me was that learning can be motivated by having an actual real world application drive the acquisition of that learning. The university students that participated in the engagement averaged $4,000 more per year in their first position than those students who didn't participate.

1983 MBA Career Planning Sessions

Part-time students returning to school to get a graduate degree in business are an interesting market. First, they are there because they want to be there and not because their parents urged them to go to school. They want value and they don't want to waste their time. I am a firm believer in aligning what you do with where you want to go and so I implemented a one-on-one planning session with every student before they even take their first course in the program. At that session I am looking for the student to answer the first two questions and then we will work together to build a program to answer the third:

- Where are you?
- Where do you want to go?
- How will you get there?

The answers give me critical information about the student's background and current job, what they want to be when they grow up (i.e., their career and professional goal) and the specific courses and sequencing they should follow to get there. The real value to the student cannot be measured. They were really motivated to complete their program because someone whom they respected took an interest in them and showed them how to achieve their educational goal and hopefully their professional goals too.

The value to me as their advisor was more than I had ever expected. Their course of study and its pre-requisite structure was my planning data. I built a course scheduling system with that data as input. For the six years that I managed the four campus graduate course schedule we never had to add or cancel a single graduate course offering.

The lesson for me was that learning can be motivated by having a short-term and a long-term career goal and the academic program to achieve it. As for me I was motivated because I was able to gather valuable course scheduling information. Clearly it was a win-win situation.

1990 Team-Driven Training

Professionals work in teams and so I believe they should be trained in teams. That is the format I have used from the beginning of my company in 1990. After several experiments I settled on 5 as the optimal team size for an effective training experience. The learning that resulted from this team-driven model sold me on the approach and I have used it ever since. It is an essential component in the WBDC Learning Environment.

Professionals work in teams and so I believe they should be trained in teams.

The lesson for me was that learning in a team-driven environment is far more productive than learning on an individual basis. Teams can create a synergy that allows them to define a solution and be self-motivating far easier than individual's working on their own.

1994 Career Agent System

The whole point behind the CareerAgent System was to create an online environment in which the individual could evaluate their current skill profile and use it to plan for additional skills acquisition activities to help them move along a career path that they designed with the help of the information codified in the CareerAgent System. CareerAgent automated the answering of the three career planning questions:

- Where are you?
- Where do you want to go?
- How will you get there?

See Figure 3.1 in Chapter 3 and the discussion that follows there for more information on CareerAgent.

The lesson for me was that career planning and professional development can be done by the individual as long as the necessary support systems are in place.

1998 Curriculum 2000

Curriculum 2000 was a bold step forward to create a BBA with a concentration in business information systems. The chair of the Information Systems Department and I designed Curriculum 2000 and implemented it in the business school at Fairfield University in Fairfield, Connecticut. The key ingredient was the full integration and participation of the business world in the curriculum. It was the first example of the WBDC Learning Environment to Business Environment linkage discusses in Chapter 4. Starting in the junior year the curriculum was fully integrated into local businesses. This was not to be confused with internship or co-op programs, which tended to be add-ons rather than integrative experiences. Student teams worked with local businesses and local businesses participated in classroom activities. Process design or improvement projects from business were brought into the classroom where they were worked on as part of class assignments and solutions presented back to the businesses. Course content was problem and project driven rather than being based on a fixed content. Most business courses were team taught and focused more on business processes than business functions. A primitive form of the Business Incubation Center (BIC) (see Chapter 3.) was employed. All of this was done in a collaborative setting with local business managers participating.

The lesson for me was that learning driven by application can be integrated into the post-secondary programs and that faculty can be productive in a team-taught model that integrates the business world. A faculty team usually consisted of four faculty members. Workshops should be used to train faculty to teach in a team-taught curriculum. Beginning in the junior year students took a 12 credit hour course each semester. The faculty teaching load was 12 credit hours so they were assigned to one 12 credit course each semester. Students formed cohorts that stayed together as a class through their junior and senior years.

2006 Blended Training/Consulting Model

I have never been a big fan of open enrollment public training programs. I don't see much of a return on investment (ROI) from such training. Rather I am a big fan of customized training for a specific client. Using a needs analysis I can zero in on the exact training needs of the client and build the training program around their existing processes and business terminology. I used that model when I formed my project management consulting practice in 1990. Later the model evolved to its current form of a blended training/consulting model which delivers even greater ROI than its predecessor. In this model I still use the needs analysis to get a clear idea of client needs and existing processes. But this workshop takes training to a new level. The purpose of the blended model is to use the training experience to teach the client team to design and improve a specific business process. The deliverable is the design and implementation plan. This is an excellent example of a process/problem-driven learning model and a major part of the entire WBDC Model concept.

The lesson for me was that training in the business environment can be done very effectively using a team approach and a problem/process-driven curriculum. Few training programs have this characteristic.

Why I Wrote This Book

As I approach the twilight years of my professional career I want to leave a record of what I have learned about career and professional development over the past 40 years. I am totally committed to making this model a reality and I will support and collaborate with anyone or any organization that shares my avocation.

Who Should Read This Book

Any professional involved in:
- workforce development programs
- training design, development and delivery
- career counseling
- entrepreneurship and business development

Specifically the following markets should be interested in this book.

- **Deans/Directors of Workforce Development.** Virtually every community college, college and university in the U.S. has a workforce development program at some stage of implementation. This book offers them some serious food for thought about what they are doing compared to what they should be doing. If what I have to say resonates with their needs, they should not only read the book but get a copy of the Guide to WBDC design and implementation.

- **Deans/Directors of Continuing Education.** The Continuing Education Department of colleges and universities is in a constant struggle to help individuals gain or regain a foothold in the economy. The need for special kinds of training is a growing market but so is the need to offer comprehensive career and professional development opportunities for their markets.

- **Members of the American Association of Community Colleges.** The primary community college market has traditionally been the adult learner. We have already talked about their plight to get into the workforce. The community college is ideally positioned to capture this market through the WBDC Model presented here.
- **Members of the National Association of Workforce Development Professionals Association.** These professionals are positioned to make the greatest contribution to the career and professional development needs of the American Worker and the American Entrepreneur. To date they are promoting the instructor-led curriculum as their major programmatic effort. That will not meet the needs described in this book but the WBDC Model will.
- **Members of the National Business Incubation Association.** The BIC component of the WBDC Model is not your father's incubation center. It is that and much more. So those organizations that have or are planning BICs might benefit from considering a broader and deeper line of business that exploits their capital investment and increases their ROI.
- **Members of the Career Colleges Association.** It would seem that these professionals should give serious consideration to the WBDC Model. It offers them a comprehensive model for reaching both the American Worker and the American Entrepreneur with an innovative program.
- **State Offices of Labor & Workforce Development.** Every state has at least one office devoted to workforce development. They need to be aware of the WBDC Model and how it might be of value across the state programs.

These offices are the collection point of all state programs related to workforce development.

- **2Yr Community Colleges.** The community college has an interesting position in the post-secondary market. Originally they were the feeder programs into 4 year programs. The thinking was that students could live at home while completing the first 2 years of a program thus keeping expenses down. Once they completed their 2 year program they could transfer their credits to a 4 year institution and complete the upper division degree requirements. Their market has expanded to include the working adult in an evening division. Their programs have expanded to include professional certification programs so the working adult could move into a career of their interests. The WBDC Model supports this trend and a community college could play a major role in providing that training.

- **4Yr Colleges & Universities.** Many colleges and universities are beginning to offer minor programs in entrepreneurship as an adjunct to any degree program. This allows the student to study a discipline of interest while preparing themselves for their future careers. The WBDC Model can expand the reach of these minor programs into business formation and professional development.

- **Vocational Education.** These programs actually begin at the high school level. Students are already integrating their senior year with career and professional development programs at nearby colleges and universities. This is often used as a motivational tool to increase the likelihood that the student will continue in high school until graduation. The WBDC Model is an excellent bridge to further education and training of the high school graduate.

- **Corporate Universities.** Corporate universities are an

interesting phenomenon and need to think about expanding their reach outside their walls and into workforce development. Their goal is to align their workforce with the needs of the organization. As an additional source of potential employees who are already aligned with their needs they might consider articulation agreements with local WBDC programs.

- **Small Business Development Centers.** New business development often involves the Small Business Development Centers (SBDC) found in local universities. Launching or partnering with WBDC programs may be a way for them to leverage their resources and support new business development.

- **Members of the National Coalition of Advanced Technology Centers.** Advanced technology centers and the WBDC Incubation Centers have a lot in common and could both benefit from alliances and articulation agreements.

- **Members of the National Council for Workforce Education.** The whole focus of this book is to introduce a disruptive innovation in workforce and business development. The current programs are dysfunctional and need to be replaced. It is no longer sufficient to offer instructor-led lecture courses and expect results. The WBDC Model is a breath of fresh air and a much needed overhaul of the education and training delivery systems.

How to Use This Book

A front to back reading is my advice to anyone thinking about implementing this workforce and business development program. As you read consider how this model can be customized and implemented in your environment. Ideally you will be a post-secondary public or private institution. Public institutions will have a few more hurdles to get past than a private institution will have. Given the state of the economy it will be challenging for both because of the need for outside funding. In addition to the new programs sponsored by government agencies at the state and national level my suggestion is to look to private industry and the many foundations and institutes they support. There is a certain bootstrapping strategy too. Chapter 7 outlines a plan for making all of this a reality. Appendix A contains an annotated list of some of these funding sources. A separate EII Publication entitled "WBDC: A Comprehensive Guide to the Design and Implementation" is available from EII Publications. Please send your interest to me at rkw@eiicorp.com.

What This Book Will Give You

The book consists of eight chapters and 2 Appendices.

There are three deliverables that this book gives you.

A PRACTICAL MODEL FOR WORKFORCE DEVELOPMENT

Monolithic instructor-led curriculum and courses are obsolete.

Models that embrace team-centric experiences and are project and problem-oriented are (or should be) replacing these obsolete approaches. The WBDC Model introduces and integrates these new models. The most significant characteristic of these models is that they fully integrate workforce development with the business environment.

A PRACTICAL MODEL FOR BUSINESS DEVELOPMENT

Project and problem oriented curriculum cannot exist on hypothetical projects and problems. They need real projects and problems to provide the motivation. Entrepreneurs and business owners are solicited to join the WBDC and submit their projects and problems to the program. Student teams will align with these projects and problems and offer their deliverables to the entrepreneurs and business owners at greatly reduced costs. It is clearly a win-win arrangement.

A GUIDE TO IMPLEMENTING THE MODEL

A guide is in preparation and will be made available for those organizations wishing to implement the WBDC Model. The guide is a 250 page three ring binder that presents a detailed expansion of Chapter 7. It offers a comprehensive design and implementation plan. An update service is also available for a small annual fee.

CHAPTER **1**

The Problem As I See It

The U.S. economy is faced with a problem whose scope is unsurpassed only by the Great Depression of the 1930s. About 10% of our workers are unemployed and lost in an economy that doesn't seem to be heading in any predictable direction and no one really knows what to do about it. The cause and effect relationships are very complex and not clearly understood so any efforts at correcting the markets can produce results very different than expected. The stimulus package is delaying the inevitable by funding public service jobs or investing in projects that have no lasting effect on unemployment or job formation. Good jobs are scarce and there is little worker confidence that if they even get hired that the job will be around for very long. Small business formation, the energy source of a sound economy, is burdened, confused and stalled. The federal government doesn't seem to side with small businesses by doing much to help in their formation. The basic tenets of our economy are rooted in capitalism and I believe that the cyclical behavior of the markets will correct the problem without the heavy hand of the government intervening. I wonder what would have happened had all of the stimulus money been awarded to small business formation and development? There must be an economist somewhere who suggested that solution!

The challenges of succeeding in the global marketplace are daunting since competition can come from anywhere and make its presence known everywhere. Small businesses don't have to sell their products and services internationally to be drawn into competing internationally. For both the worker and the entrepreneur the impact of technology can be devastating too. Technology is not geographically constrained and is permanently replacing jobs every day. Yet creatively leveraging technology is a critical component of most sustainable businesses. The dilemma is truly a challenge for our most creative minds.

... Competition can come from anywhere and make its presence known everywhere.

Drawing a circle around this problem is a challenge. If the circle is too big, the resulting solution (if you can find one) may be too general to be implemented or to have any sustainable market value. Should we try to solve world hunger or just hunger in Haiti? Draw the circle too small and you may end up solving a problem that nobody really cares about – again producing no value. The problem scope that I have settled on in this book is broad.

The workforce and business development problem I am going to describe is complex, filled with uncertainty and constantly changing. The casual observer might be tempted to think there are really two problems here. One problem is the worker who seems to be cut off and adrift from the mainstream economy and needs an entry strategy and a sustainable career. The other is the entrepreneur who is also adrift and not sure how to proceed with the formation of a business that has sustainability too. In fact what

I see are two problems that are inextricably linked to one another and I will show you how. This problem can be solved with a single solution. That is the position I am taking in this book. I am taking a holistic view of the problem but not without offering an adaptive, robust and disruptive innovation to solve it. I believe that the disruptive innovation introduced in this book is just the solution we should be implementing. And while it is only the first step at a lasting solution it is the correct first step. And it is the best first step!

The problem is best described in terms of the three major components that comprise it and that are also part of the solution. Those components are: the Learning Environment, The Business Environment, and the Student Environment. They are discussed in the next three sections.

The Learning Environment

Whatever form the workforce and business development solution takes it must be supported by a comprehensive education and training component beginning at the high school level and extending through the community college, college and university system and even to the university graduate level. We live in the information age and the professions to target will be those that creatively integrate information technology into every type and size of business and physical process. Even in these hard times there are many such opportunities just waiting to be found and exploited. All that is needed is some creative, out-of-the-box thinking. The proposed WBDC Model is structured so that these opportunities can be discovered and programs to exploit them developed.

One of the major obstacles that I see is the need to extend the reach of education beyond the mere presentation of concepts, theory and principles to embrace real world applications. This is the monolithic delivery model that dates to the Industrial Revolution. It is based on a weak assumption that if the student simply pays attention, they will learn. For some that is true for most it is not. These models are now obsolete and must give way to team-driven project-based learning models. For some faculty this will be a welcomed challenge that they will quickly accept. For others this will be very difficult to accept and adopt. For a few, this change will not be possible. The looming question is what is the role of the faculty in this new paradigm? In addition to being a subject matter expert I see them taking on a second and more important role – that of facilitator. It is unrealistic to expect a single faculty member to have the experience and knowledge base that would be needed to contribute the complete content to every project situation that might arise. They need to have the awareness of where a student can go to get the information and specific learning experience they need. Clearly the internet is indispensable. The key to supporting these new programs will be the faculty's ability to integrate the business environment into their learning environment and vice versa. The solution must provide the needed infrastructure, resources and support so the faculty and their students can do this effectively.

The monolithic delivery model must give way to team-driven project-based learning models.

By way of summary the problems that I have observed with our high school and post-secondary educational delivery systems are many and the conversion to a team-driven project-centric model will be difficult. I've listed those problems below in no particular order and without the need for further comment. The book will impact many of them. In some cases these are just a restatement of commonly accepted problems.

- **Drop-out rates are too high – graduation rates are**

too low. I have personally experienced students with good potential completely turned off by the system and dropping out before program completion. They can succeed but the motivation is not there. That can be corrected. I have also worked with a number of high school students and high school graduates who just needed the motivation and an achievable plan in order to succeed. One that I will never forget is a first generation American student from a welfare family who had an aptitude for, and interest in, computers. Let's call him Pedro. Pedro got involved in a magnet school that gave him a chance to develop that interest. To make a long story short, he got really turned on, won a programming contest and was awarded with a 4 year all expenses paid scholarship to MIT to study computer science! How many Pedros are out there waiting to get a chance but are not likely to because of the present system? What a terrible loss to society. How sad.

- **Monolithic delivery systems are broken and must be replaced.** There has not been a major overhaul of the educational delivery system since the beginning of the industrial revolution - the internet excepted. Many students need the motivation of an application in order to learn. Monolithic delivery systems cannot do that effectively. They cannot be incrementally improved. They have to be totally replaced.

- **We need disruptive innovation not incremental change.** Incremental changes are like automating a decision model. In most cases the result is to make bad decisions faster but not better. For some educators a disruptive innovation will be very difficult since they are so entrenched in and protective of their ways. Any change will be difficult.

- **Programs have failed to meaningfully integrate business partners.** Internships and co-operative education programs are largely failures. I once met an engineering co-op student whose co-op job was working the counter at MacDonald's. How sad. Students often have no motivation to learn in the absence of application. If they knew of the importance by seeing the application before the learning takes place, the problem becomes mute. So the better model is to have the need and then be tasked with figuring out what has to be learned to satisfy the need. The ancient Chinese proverb says it most effectively:

Tell me and I'll forget.
Show me and I may remember.
Involve me and I'll understand.
                                    Confucius 450BC

- **Balancing cost/SCH against program requirements presents competing alternatives.** The post-secondary educational business model is dominated by competing alternatives and limited financial and faculty resources. Do you offer a course that is undersubscribed because it is required in the degree program and there are only a few students who need it now or do you offer a very popular general elective? You can't do both. Usually a substitute or an independent study alternative is offered in place of the required course. How sad. If the course had a broader market reach, like to collaborating institutions, the problem might be mitigated. Few institutions have programs in place to reach beyond their own students for registrations. There are opportunities for collaboration to the benefit of both institutions but few institutions are creative enough to take advantage of them.

- **There are constant pressures on marketing and recruiting programs to generate applications.** You can only carve the market up into so many pieces and sooner or later the cost per application increases. As stated above a collaborative may resolve the problem. The program portfolios of most post-secondary institutions are over-extended. What they are doing is equivalent to trying to put 5 lbs of dirt in a 3 lb bag. I believe that a strategy of collaboration rather than competition across the market is a better strategy but very few institutions have such strategies. How sad.

- **Absorb growth without increasing the budget.** Post-secondary education has had budget problems for years and despite the growing and shifting markets they have to serve they often get less than adequate resources to support their market. Part of the problem is the entitlements that take a larger and larger percentage of the budget. That leaves little for program development. In business that situation is like laying off the sales staff due to budget problems. How sad.

- **Need for new revenue generating programs.** Traditional programs are often run at breakeven numbers or worse. New programs have to become self-supporting quickly and even bootstrap themselves financially. Higher education is a business! Run it like a business!

- **Meeting program/course registration targets.** So many degree programs are not running at breakeven levels. Enrolments are not sufficient to allow undersubscribed courses to be run. Substitutions (i.e., independent study) often have to be used to generate the required credit hours in many degree programs. It has been so very difficult for academic administrators to face the problem and make the correct business decisions. Across the board budget cuts of x% tend to lower all the boats in the harbor and everyone suffers. How sad.

The Business Environment

We all know that local, state, national and global business communities are in the financial tank and our best minds have only been able to guess at the steps that might be taken to correct the situation. Republicans are of the opinion that capitalism and the free market will correct the situation and no government intervention is needed or advised. Conversely Democrats believe that only with heavy-handed government intervention and massive financial support will the economy recover. It may be too optimistic to expect that the multiplier effect of an injected dollar will create lasting economic stimulus but that is all our politicians have to rely on. By contrast the two major political parties are deeply divided over the role of government and what should be done. There seems to be no end to the continuing partisan battles and they aren't delivering much in the way of meaningful progress. In the meanwhile small businesses are adrift and workers sit by and watch in seemingly helpless desperation. The problem is clouded by the usual political jockeying. Politicians seem hell bent on getting re-elected and will do whatever they think will promote their goals. Whatever the opinions of their constituency might be they are not being taken into account. The Democrats have dug themselves a deep hole over the health care bill and we have the rise of the Tea Party which is hell-bent on seeing the Senate and House stripped of all Democrats. Maybe their success is the best thing to have happen. Times are changing. Economists have proposed as many solutions as there are economists. To everyone's credit very few living persons have ever encountered as complex a situation as the one we are now faced with. So there is no historical record or experiences to draw upon. The one lesson that seems to be emerging from all of this is that the worker and the entrepreneur are on their own. Corporate loyalty to the worker is only an historical memory and there is little evidence that it will ever return. In the absence of consistent government support the entrepreneur must fend for themselves.

The situation has been exacerbated by the continued weakening of our manufacturing base and the trend among US companies to move many technology-based jobs off-shore to reduce costs. Global competition has become pervasive even for companies who only sell their products and services in the U.S. Their competitors can be anywhere. Companies are recruiting internationally to fill technology jobs based in the U.S. when there are qualified US citizens available. Companies assume there will be salary savings resulting from hiring an international candidate as compared to a US citizen.

To further add to the problem jobs are continuously being replaced by the unrelenting advance of technology and are lost forever. Technology has simultaneously become both a weapon and a shield.

Virtually every business has become a participant in the global economy. And by that I don't mean that they sell their products and services internationally. Rather our businesses are competing with global businesses who sell the same products and services that they do and they do it in our own backyards and often at less cost to the customer. The fishing industry is a prime example. Staffing call centers is another.

To be successful in these global markets businesses must

exploit technology to protect and advance their market position. Simply using technology to enhance operations or increase services isn't sufficient if it can easily be replicated and encourage the growth of new competitors whether they be in Mumbai or Moscow. Businesses need to find creative and innovative ways to apply technology to product/service offerings in order to increase revenue avoid costs or improve services (IRACIS). To be successful businesses will have to create new jobs or redefine existing jobs by leveraging technology in new and innovative ways and to do it so the competition can't easily replicate it. To be sustainable these jobs need to be designed so they cannot easily be outsourced. Not all new jobs will have this property and that needs to be taken into account as part of new business formation. The workforce must be prepared to fill these new jobs and I believe that will only happen by implementing a disruptive innovation in the form of some WBDC Model such as the one discussed in this book.

Regardless of the state of the economy when this book is published the worker and the entrepreneur will have gotten the message. They must develop a defensive strategy and also take the offensive so that such a catastrophe will never again befall them.

In the midst of all this doom and gloom the WBDC will carve out its place for students, workers, entrepreneurs, and business owners to seek and find the special kinds of proactive support they need to survive and thrive despite the odds.

THE STUDENT ENVIRONMENT

The plight of the student is every bit as complex and challenging as the entrepreneur and business owner. The use of the WBDC Model by the student is a lifelong relationship. To begin the discussion of the problem facing students, educators, trainers and organizations as they look for practical and realistic models that can meet the lifelong career and professional development needs of the student there are six different markets to consider. Those six markets are:

- **High School Sophomores.** For the purposes of this book the problem begins at the high school level. It probably has its roots much earlier in the elementary level but that is out of scope for this book. There are a significant number of technology-gifted high school students who have completed their sophomore year and who are not planning or able to attend a four year college or university at this time. In many cases they are not even motivated to study or to graduate from high school. The high dropout rate reflects this lack of motivation. These students are a valuable asset and it is criminal to lose them to the system. For many all that is missing is the motivation and the means to participate. The WBDC must embrace these students and motivate them to complete their high school education while at the same time acquiring the skills, competencies and WBDC support to launch their own technology-based career or to mainstream into existing businesses. I'm constantly reminded of the success of the Washington High School magnet school program in Milwaukee and I know these lost and misguided students can be helped. When these students graduate from high school and complete their WBDC program they may be ready to take advantage of several articulation agreements and mainstream directly into a community college or university program at any one of several participating institutions. Hopefully they will continue to include the WBDC as part of their career growth and professional development.

- **High School Graduates.** Technology-gifted high school graduates who are not planning or are currently unable to attend a four-year college or university make up another significant market. Many will have to work to support their families. Many communities have students who are first generation Americans and post-secondary education may not be a priority for them. They often have no role models either. Further education might not be on their radar screen. If they have a parent role model, they are fortunate but many do not. The WBDC will provide the mentors and is designed to motivate these students and help them plan their careers and professional development and then acquire the skills, competencies and support to launch their own businesses or to mainstream into existing businesses. Articulation agreements with colleges and universities may provide other avenues to further their career and professional development through further post-secondary education.

- **Traditional College and University Students.** Many new business ideas arise among the traditional college and university student body. Bill gates founder of Microsoft and Michael Dell founder of Dell are two examples that we all know about. A few universities are beginning to implement entrepreneurship minors in many of their degree programs. These minors could have a strategic, tactical or operational focus and there are several examples at the community college, college and university levels.

- **Un-experienced Adult Workers.** Adults who have never been employed or whose work experience is so old that they must prepare themselves for a new technology-based career and to be able to compete in a technology-based business world form another sizeable market. Many of these people may have undergraduate and even graduate degrees but are technically challenged. Through the WBDC BIC this group may choose to use the WBDC for exploring business ideas with a view to new business formation.

- **Experienced Adult Workers.** This group consists of displaced, unemployed or underemployed workers who have lost their jobs for a variety of reasons outside their control. Most will be the victims of outsourcing, technology replacing their jobs and down-sizing. Their goal is to get back into the economy and stay there. For many that will mean forming their own business but that will be a challenge. Too many business ideas wither and die because of technology despite the fact that technology may also be the competitive advantage. Others will fail because of the incursion of international competitors.

- **Recently Discharged Military.** The men and women who have served us so courageously and are now in the job market need special assistance and retraining to be competitive in the current and future job market. Part of the challenge is identifying their transferrable skills, adding to them very selectively, and preparing them to enter the civilian workforce.

These six markets span the entire workforce from first time entrants, to mid-career and returning workers. While this may seem like an aggressive target market for a single WBDC it turns out that these six markets are complementary of one another in the career-spanning WBDC Model. Because the WBDC Model embraces individuals at all stages of their careers it provides opportunities for a person to grow professionally under the guidance and lifelong advice of the WBDC regardless of where they are in their careers. As members of a student team they collectively create a synergy that will prove valuable to the projects submitted by member businesses.

The book presents a single innovative WBDC Model to serve all six markets concurrently. The Model is described in detail and a development plan presented. As I began sharing the Model with my clients it caught the attention of two higher education institutions. Both institutions expressed high interest and indicated their intention to begin implementing the Model in 2010. I will maintain my involvement with them and if there is any progress to report on their implementing the WBDC Model, I will document their experiences for you.

The book offers a unique approach to workforce development that has every indication of success. As you will see it is a comprehensive approach to workforce development and is robust so that its application across several disciplines is an application waiting to happen.

At the time of this writing the Model is conceptual but I believe very practical. To support its implementation I have written a companion book that is a planning and implementation guide for adopters. That guide is available from EII Publications at rkw@eiicorp.com.

The most recent unemployment percentages hover just under 10 percent – the worst since the Great Depression. When you add the number of workers who have given up looking or have accepted part-time and lower level positions than they are qualified for, the number of displaced workers is staggering. Some have estimated that the real unemployment percentage is 17%. I think that is optimistic. This situation is clearly unacceptable. But the problem is exacerbated by a complex and constantly changing business environment as discussed above. Technological advances, global markets and the transition to off-shore development are contributing factors. The result of this changing business environment is that a significant percentage of technology-related workers are now unemployed and looking, unemployed and not looking at all (they have just given up), displaced, and underemployed. In addition there are a significant number of adults who have never or not recently been employed but financial circumstances now require them to become the second breadwinner in their family. They are particularly challenged because many of these people are technically unqualified to compete for jobs in today's high-tech market. Many of these peoples will be looking for points of entry into a business world that expects varying degrees of technical knowledge and skill. Where are these points of entry, what skills are needed, and how are these skills acquired are all questions needing answers.

Then there is the underlying question of sustainability of these positions too. Do these positions have any staying power? Most of the jobs created by the stimulus package will not. They are simply projects that will end and the jobs associated with the project will disappear. Others feel compelled to start some type of family-owned business but are not sure what business or how to proceed. One thing that is common to all of these people is the need to identify and then prepare themselves for positions that cannot be easily outsourced. Many of the more secure positions that offer growth opportunities will have some form of technology component. And that is a double-edged sword as we know.

Still others could put forth an argument that technology is also part of the problem and the solution would be to launch a position or start a business that does not and cannot use technology for competitive advantage. Service oriented businesses that require physical contact with the customer would be an example (i.e., plumbers, electricians, painters, bakers, taxi drivers, public servants, etc.). The question about long-term career growth, if there is any, becomes an issue with these types of positions.

THE PROBLEM

So the four target markets for a solution are workers, entrepreneurs, faculty, and business owners. The holistic solution must deliver to the needs of all four markets. To meet these needs effectively will require a disruptive innovation like the WBDC Model.

Worker needs

Worker needs include the following:
- Identification of sustainable career opportunities
- A program of study to give them the best chance of entry into and growth within their chosen career
- Counseling and mentoring to help them make good career decisions
- The opportunity to apply their new knowledge to meaningful experiences
- An understanding of the business environment
- Practice, practice and more practice

Entrepreneur needs

Entrepreneurs needs include the following:
- Safe harbor to objectively and intelligently explore ideas
- Business planning support
- Strategic planning for creating a sustainable business
- Planning to leverage technology creatively
- Start-up business services
- Ongoing support of their early business needs

Faculty needs

Faculty needs include the following:
- A plan to incorporate project/problem-driven curriculum models

- A source of realistic or actual projects and problems
- A referral source of information for student teams
- Collaborative partnerships with entrepreneurs and business owners
- Resources and administrative support

Business owner needs

Business owner needs include the following:
- New business formation
- Process design
- Process improvement
- Problem solving
- Staff development

PUTTING IT ALL TOGETHER

To summarize the problem let's look at it from two perspectives: the worker and the business owner. For the worker the problem is:
- they are not employed at the level their skills and experiences have prepared them for
- they fear losing their next job to an off shore company or to technology
- they do not have a career plan and don't know where to turn for help
- they have been thinking about starting their own company but need help

For the business owner the problem is:
- they have an idea for a new business and don't know where to turn for help
- they have lost market share and can't afford the professional help they need to identify and make changes
- the performance of their business is less than acceptable and they can't afford the professional help they need to correct the process problems

So the scope of the problem is now defined. Its solution isn't going to come in the form of tweaking what is already in practice. Its solution will be a complete replacement of worn out and ineffective models with a new and innovative model. Because I am advocating a wholesale replacement I am asking people to step outside of their comfort zone and with an open mind adopt a different approach. That will be asking them to be courageous but that is what it will take to solve this complex problem.

CHAPTER 2

Solution Requirements

Before we take up the topic of the solution in Chapter 3 I think it is worthwhile to describe the requirements that an acceptable solution must meet. That will give us a benchmark against which the solution presented in Chapter 3 can be assessed.

Requirements gathering and documentation is a challenge even in the best of circumstances. Since I am proposing a disruptive innovation I am treading into unknown waters and what the final solution will look like can only be guessed at the outset. That is the nature of every disruptive innovation and the WBDC Model is no different. But I do feel that enough of the solution requirements have been identified so that the initial solution template will be close to the final solution template. So we have to understand that the project to design and implement a solution that meets specific institutional requirements is a best guess at least initially. The impact of time, discoveries and application experiences will be integrated into the initial solution and it will continuously converge to a better solution. This cycle of learning will continue into perpetuity. So your version of the WBDC Model will never be finished. WBDC Model implementation is really a continuous process improvement project and must be approached as such.

I feel privileged to be here at this point in time and to have an opportunity to contribute by taking the first step. What I propose in Chapter 3 is the best approach that my experiences and understanding of the problem suggests. You have to think of the proposed WBDC Model as a template that must be customized to meet specific needs. Time and practice will suggest changes to improve the effectiveness of the solution and those changes will be made.

TRENDS THAT WILL AFFECT EVERY SOLUTION

There are a number of internal and external trends at play in workforce and business development that are obvious and have to be accommodated for any solution to be effective and sustainable. To be effective and stainable the solution must be robust enough to be adaptable to changing conditions in society, the business environment and worker needs and to a certain extent it must anticipate those changing conditions. The robustness will come from constantly monitoring the environment and adjusting accordingly. I have designed my solution to be a solution for all times. It is sustainable. I know that it will change to accommodate changing conditions and needs but I can't presume to guess what those changes might be. We know some of them and they are described below. What I do know is that the solution must accommodate certain trends that have already established themselves. The trends that any solution must accommodate include the following:

- **Global perspective** - While the market for a new business idea may not include customers globally its global competitors might and that puts offshore businesses in direct competition for your domestic customers. So any new business development plan must have a global perspective and offer sustainability. In some cases the business must be located near the customer but in other cases the business idea must be creative and leveraged by technology in a way that is not easily replicated otherwise it will not present barriers to entry for new competition and therefore will not be protected. Culture can be a barrier to entry for those of a different culture. The written word can also be a barrier. Those for whom English is a second language often have a difficult time communicating clearly and effectively and so the communications industry is often safe from the incursion of businesses for which there are significant cultural differences and for which English is a second language.

- *Information technology and business function integration* - Many would argue that technology is going the way of the telephone and will soon lose its identity as a solution provider and become an infrastructure provider. For the worker and the business owner this translates into their having more than a superfluous understanding of technology as applied to their businesses. They must know enough about technology to be able to integrate it into their business formation and development projects.

- *Pervasive technology* - Technology is pervasive in almost every business. The challenge is to creatively leverage technology for sustainable competitive advantage. I remember when greeting cards were first equipped with microchips loaded with a song or a verbal message. People used to say: "Is that a greeting card that thinks it is a computer or is it a computer that thinks it is a greeting card?" I guess it is worth wondering how many of these applications of technology are waiting for a creative mind to exploit. The social networking rage is a good example. While the creative use of technology can be a barrier to entry it has also resulted in the loss of several jobs that are now permanently lost.

- *Systemic business problem-solving* - Solving business problems can be the most challenging of any creative process you will need. If you doubt me, check out Southbeach Modeler (www.southbeachinc.com) and P-TRIZ (www.docstoc.com/docs/1164883/Smith--Introducing-P-TRIZ). These recent entrants into the problem solver's toolkit will show you just how creative the problem solving process has become. To be effective in developing job entry skills the student must understand business problem solving from both a knowledge basis and an experience basis. This requires the use of contemporary tools, templates and processes as well as a partnership between the Learning Environment and the Business Environment.

- *Increased flexibility of educational delivery systems* - The monolithic delivery systems are dinosaurs held over from the Industrial Age. They are rigid and unbending. The few incremental changes that have been made haven't made a significant difference as measured by any performance metrics that are commonly used. A total replacement with a project/problem-driven approach is needed. That is a step outside of the comfort zone of most faculties and begs of a partnership between the Business Environment and the Learning Environment. The Business Environment must supply the Learning Environment with projects and problems that will be the platform for learning. The application must precede the knowledge needed to do the application. Case studies reverse the timeline and ask the student to apply what they have learned to a real or hypothetical situation. That approach offers little in the way of motivation to the learner to learn and so that isn't the answer.

- *Increased continuing education* - There will always be

a demand for continuing professional and vocational education. An effective solution will recognize this and offer an educational delivery system that can accommodate the changing needs of the worker. For businesses they will have to provide the projects and problems that will be the platforms on which their employees can learn. That will require a partnership between the Learning Environment and the Business Environment heretofore unexplored. The project/problem-driven learning model has never been used for continuing education as far as I know. But that has to change to deliver effective training. I'm expecting the trend will evolve to an online delivery model for continuing education for the employed worker. This will be very specialized but in time will generate a broad and deep online inventory. That inventory will consist of a number of pre-programmed learning modules.

- *Use of the internet as a learning and marketing tool* - Clayton Christensen ("Disrupting Class: How Disruptive Innovation will Change the Way the World Learns," New York, NY: McGraw-Hill, 2008) has described how the internet has ushered in a disruptive innovation in K-12 education delivery. No one would argue that. But the internet is poised and waiting for the next foot to drop. It is a limitless source of new and relevant information regardless of the topic. In effect the internet is evolving into the text book of all text books. The biggest problem is summarizing information to the knowledge level. Doing a search and getting 500,000 responses is nice but falls far short of the mark for the student. I use the internet quite extensively and often get frustrated by the information overload and knowledge scarcity.

Somewhere out there is an untapped business opportunity to develop a data warehouse application that takes information as the raw material and produces knowledge as the product.

As a marketing tool the internet greatly expands the market for products and services. The internet expands your market reach to the entire world. Even if the U.S. market for your product is small, the global market isn't.

- *Partnering with business* - Learning without the opportunity to integrate experience and application isn't really a very effective and lasting learning experience. Businesses are the only source of real applications and must be part of any educational delivery system that claims to be effective. Internships and co-operative education have been around for decades but the results from those programs have been mixed. Integrating those programs into the learning experience is weak at best. Both models are after the fact and hope the student will link their knowledge with application. A more efficient and effective approach is for learning to be driven by the knowledge and practice needs of the application. That model is only beginning to be seen at the high school level and in only a few colleges and universities. It has become quite common in vocational education curricula.
- *Increased outsourcing* - As U.S. businesses look for cost avoidance opportunities the shifting of the labor part of a product to an off-shore interest has been a cost saving strategy. As more firms in a particular market utilize that strategy the advantage within their market lessens. But

outsourcing remains a factor in the design of new or improved products and services. The creation of businesses that cannot easily be outsourced in order to create a cost advantage is still a good strategy for the sustainability of your business idea. The WBDC programs, curriculum and courses must pay close attention to developing business ideas that are sustainable.

- **Decreased traditional age students** - Learning has become a lifelong journey for everyone. It never stops and the need for it never stops either. So a comprehensive learning environment must adapt to the changing needs of every worker and entrepreneur. It is already shifting from an education model to a training model.
- **Less manufacturing** - Manufacturing has moved to countries whose labor costs are lower and the manufacturing jobs would seem to be gone forever. The base of our economy has moved from manufacturing products to processing data in order to manufacture information and knowledge products. Robotics and other manufacturing applications of technology may return some equality to the competitiveness of the U.S. economy.
- **Sustainability** - New businesses create staying power by being businesses that cannot be easily replicated here or abroad. Other than a location requirement, that means creatively using technology to define and operate your business. The solution must offer the entrepreneur the support they need to test their idea for sustainability.

WBDC SOLUTION REQUIREMENTS

So with all of the above taken into consideration here are the requirements that must be met to deliver a solution that accommodates and even takes advantage of the above trends. I've collected the solution requirements into four groups: Learning

Environment requirements, Business Environment requirements, Worker Environment requirements and BIC requirements. The entire Requirements Breakdown Structure (RBS) architecture is shown in Figure 2.1. The level of decomposition needed to describe each requirement is usually different for each requirement. Some requirements are very simply stated and don't require much decomposition to be fully understood. Others are quite complex and require considerable decomposition to be clearly and completely defined. The RBS is not an exercise in symmetry. It is an exercise to describe a solution. As much as is known about each requirement can be displayed using up to 5 levels of decomposition as shown in Figure 2.1.

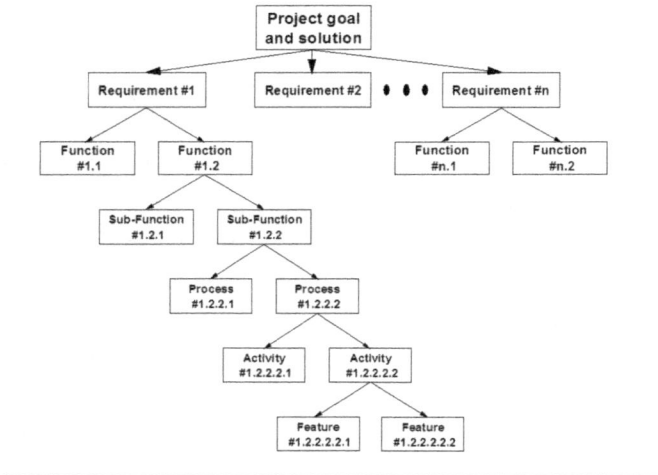

Figure 2.1: RBS Template Decomposition Structure

Describing the complete requirements is a best guess and it must be viewed in that light. For the sake of space I show only a few levels of decomposition for all but the BIC requirements. The BIC is so unique that I thought it would be beneficial to describe it in as much detail as I can. You still have to keep in mind that I am only presenting a template. You will have to customize it to your situation. For the BIC part of the RBS I show the full decomposition at least as far as it is known at this time.

- **Requirement** - There are four different types of requirements: functional, non-functional, global, and product/project constraints.
- **Function** - specifies what the product or service must do or a property it must have in order to do what it must do
- **Sub-function** - a necessary and sufficient decomposition of a major function into the several smaller scoped functions that together cover all requirements of the function
- **Process** - one or more physical processes that describe what must take place in order for the function or sub-function to be completed
- **Activity** - one or more physical steps that must occur in order for the process to be completed
- **Feature** - a physical characteristic of the activity

The hierarchical decomposition works well on a whiteboard but not in printed format. For that I prefer an indented outline format that is given in the BIC example below. The same information can be displayed in both formats. I have found the format in Figure 2.1 to be particularly effective for group presentation on large whiteboards. Using that format the entire RBS will always be visible and clear. The indented outline format works best in printed format because it is economic in its use of white space. Because we are dealing with a disruptive innovation we can only take the best guess at the RBS for the implementing institution and that is what I have done. Once the WBDC is implemented and there has been some practice using it, it will change. That is for certain.

For the WBDC Model a complete RBS (that is as complete as we know it to be at the outset of the project) is not instructive to put in this book. But I will include the portion of the RBS that describes the BIC. I do that for good reason. First, it is different than your father's incubation center and I want you to understand how I see it because it is so fundamental to any WBDC Model design and implementation. Second, I want you to have some examples of an actual decomposition.

Learning Environment Requirements

The WBDC Learning Environment is totally different than the vision you might have of it. First of all, it would be unreasonable to expect that the instructor assigned to "teach" you will have sufficient knowledge of every topic that might come up in your project. But it would not be unreasonable to expect the instructor to be able to point you in the direction where you might find the information needed to "learn" about a topic. "Teaching" and "learning" will take on different meanings than most instructors are comfortable with. "Teaching" and "learning" is done both by the instructor and the student. That is the nature of a team-driven Learning Environment." The instructor is just as much a person who delivers knowledge as they are a facilitator who directs students to sources of knowledge. With that thought in mind let us look at the requirements of the contemporary Learning Environment.

Be based on the concept of a "classroom without walls"

The monolithic learning models that grew up in the industrial age are obsolete. Clayton Christensen (*Disrupting Class: How Disruptive Innovation Will Change the Way the World Learns*, New York, NY: McGraw Hill, 2008, ISBN 978-0-07-159206-2) writes eloquently about the changes that are taking place and will take place as the internet invades the 21st century classroom. Thanks to technology the learning environment is now a classroom without walls. Information exists in all sorts of forms and locations and information is the beginning of the learning process. So in the WBDC Model learning can take place anywhere there is information that must be assimilated by the student team. Those learning places can be any or all of the

following:
- Classroom
- Faculty
- Guest lectures
- Interviews
- Books
- Internet
- Library
- Business partner meeting
- Professional society meeting
- Student team meetings
- Conference
- Government hearings

... learning can take place anywhere there is information that must be assimilated by the student team

Include learn-to-work and work-to-learn components

This statement really emphasizes the need for integration of the learning environment and the business environment. In order to learn, many learners excel when they learn in the same environment where their learning will be applied. There are two benefits that accrue to the student team from this scenario.

- The first is that they learn to work.

    For some students their first exposure to the business environment can be daunting. They have to learn how to get along with co-workers whose habits and practices may be strange or even unacceptable to them. They need to learn interpersonal and personal behaviors that the typical book learning doesn't offer. It would be best if they could learn to work in a non-threatening environment. The WBDC Model must offer that environment.

The team structure provides that opportunity as does observing and participating in activities with member businesses.

- The second is that they work in order to learn.

Knowledge does not mean skill and not until the knowledge is practiced can a student say they have learned something. In other words, the knowledge must be practiced to be really learned and mastered.

Utilize team-driven learning & discovery models

Almost everyone works in teams so it would seem logical that they learn in teams. Not only is learning and discovery more efficient in a team structure but the side benefit is that you learn how to be a team member. There is no course that can teach you how to work in a team. The only way to learn it is to do it. So the curriculum is designed around a team-driven model. That is one benefit to a student but there is a second of equal importance.

The second is the synergy that follows from a team approach. One person can learn something of value to the team and share that learning with the team. Ideas that are shared can multiply whereas an idea that is not shared is simply an idea. The team can motivate itself to learn and master some knowledge or skill that would not be possible on an individual basis. Not wanting to let your team down by shirking a responsibility is not likely to happen in a team-driven learning model. You want to be seen as a contributing member of your team and so you will complete your assignments completely and in a timely manner for the sake of your team.

Almost everyone works in teams so it would seem logical that they learn in teams.

Be based on a problem/project-centric learning model

Being motivated to learn and experiencing the application of learning is best done by immersing the learning in an application. So the curriculum is based on a problem/project centric model.

The key to adopting a problem/project-centric learning model is to have a source of problems and projects. But understand that these are not after the fact problems or projects. That is, the learning model is not one where you acquire the knowledge and then go looking for an application of the knowledge. Rather it works the other way around. That is, you begin with a problem or project and then go looking for the knowledge needed to solve the problem or conduct the project. Those institutions that are trying to implement this model depend heavily on hypothetical projects. While that may work at the high school level there is a better alternative at the post-secondary level and that is to use real projects.

For the instructor this is a big challenge. The internet can be a resource as other instructors have faced the same challenge and they have submitted problems and projects that might be adaptable to an instructor's specific need. Alternatively, they might turn to a resource available through the WBDC and that is the approach I envision here.

Provide a comprehensive resource for career and professional development

The WBDC intends to be a lifelong resource for the worker. It must offer a beginning to end service that provides them all of the advice and support services they will need across their entire career. That begins with an assessment of their current skills and job descriptions, continues with a description of their short-term and long-term career goals, continues with a plan to achieve their career goals and finally offers monitoring and coaching services to support career attainment. All of this is imbedded in an education, training and experience environment.

Offer growth opportunities to every worker

Getting your first start as a worker or owner of a new business are daunting tasks filled with risk and failure. Getting started is hard enough but developing the staying power is even harder. The WBDC has to build credibility to serve both of these interdependent markets.

For the worker the solution needs to stay close to them, their needs, their failures and their successes. Those needs will differ from worker to worker. So the solution must be adaptive to the worker. Nothing less will suffice.

Business Environment Requirements

I've long held the opinion that the business environment is an untapped resource for learning. For example, a project consists of a number of interdependent tasks that must all be completed in order for the project to be complete. Some of these tasks are not critical to the project schedule and therein is the key to using projects for on-the-job training. Rather than assigning an experienced person to a task whose skill requirements they have long since mastered, why not assign a more junior person who has not mastered all of the skill requirements needed to complete the task. They know they do not have all of the skills and so do you. But in making that assignment you have agreed to provide the person with the opportunity to learn the needed skills and then practice them on their assigned task. They will probably take a bit longer to complete the task than the experienced person but the project schedule can accommodate that.

Support entrepreneurship and business formation

For the entrepreneur the WBDC Model offers a beginning to end service that provides them all of the advice and support services they will need beginning with a business validation study for their new business idea to test its feasibility or adjust it so that from a strategic perspective it is a sustainable and feasible venture.

That service continues by supplying all of the research needed to create and submit a defensible business plan. For the embryonic business process design and improvement projects are frequent and support will be available for that too.

Careers in other people's organizations are not a popular choice. How often do you think people choose organizations? I think they choose organizations for their careers when they don't see any practical or attainable alternatives. Organizations have a lot of negative baggage and it will be difficult to overcome. Lack of employer commitment and loyalty to the worker lead the list.

Rekindle the American "spirit of innovation"

Our nation was built on the backs of courageous people who were willing to put themselves in harm's way to pursue the idea they believed in. We have seen this happen during a student's undergraduate education (Bill Gates and Michael Dell). I have personally observed it during a student's high school years. There are so many high school students with a strong attraction to and understanding of technology that only need a slight push and a pull to become successful entrepreneurs. The WBDC will provide that structure. Robotics and vision systems have taken over many manufacturing processes especially in the automobile industry but these have been easily replicated. But there are still many undiscovered applications just waiting for a creative mind!

Student/Worker Environment Requirements

The Student/Worker will have a number of requirements that are either part of the Learning Environment or available through BIC projects.

Exploit technology for sustainable business and job creation

Technology and technological advancement are both good and bad.

They are good because they create opportunities to leverage a business or launch a new business. All that is needed is a dose of creativity and some sweat equity. It is good because the cost of entry is low. Many businesses are virtual businesses and operate exclusively on the internet. All you need is an idea that is sustainable and not easily replicated by the guy next door or across the globe.

They are bad because they tend to displace workers or at least create business outsourcing or off-shore development. It is bad because it doesn't have any borders. A dining room table in a Mumbai home may be the corporate offices of your major competitor. Can your customers discern the differences between what you offer and what your Mumbai competitor offers? If all business is conducted over the internet, probably not. We've all had experiences with call centers where the service is provided by a person somewhere outside of the U.S.

Concurrently meet the needs of business and academe

Businesses have to leverage technology to become competitive and to remain competitive. Continuous process improvement must in fact be continuous. Businesses must always be on the lookout for opportunities to leverage technology to increase revenues, avoid costs, and improve service (IRACIS).

Post-secondary educational institutions have to leverage technology to become competitive and to remain competitive. The internet is a growing factor in the Learning Environment and the WBDC curriculum must embrace its use.

Continuously align to changing markets and competition

The world doesn't stand still just because your company wants to be a successful player. What worked yesterday may not work today because a competitor upstaged you with a bigger, better or more cost effective product or service. Product and service lifetimes are shorter than they used to be. That means your next release has probably got to be in the pipeline even before your current release is in the market.

Not only that but you are a global business and that has little to do with where you sell your product or service. Your competitors are global and they play in your markets. So you are forced to compete globally.

Be financially sound and create social value

Let's face it, the WBDC is a business. First of all, it has to be able to support itself financially. In addition it has to create social value.

Be flexible

If the WBDC is to create social value others must be able to adapt it to their situation and implement it. The WBDC Model must travel to other environments, countries, and even cultures. That is the motivation behind the Guide. One of the major problems of current educational delivery systems is that they are fixed and do not accommodate the different learning styles of their students. The WBDC Model must adapt and embrace adult learning theory especially as applied to the six different student markets. These six markets have different expectations with respect to how they will learn and the WBDC must be prepared to deliver to all of them.

- Scalable

   The WBDC Model must scale well. It is a community resource. It must adapt to very small and restricted markets as well as very large and comprehensive markets. Entrepreneurship and business development can be very generally targeted and the WBDC should accommodate that. On the other hand, a small isolated community that is supported by one or two industries presents a very different kind of problem for the WBDC. Only businesses directly related to or suppliers of the major industries present make sense. The usual service related businesses that support the community are always presenting opportunities.

- Replicable
  The WBDC Model must travel well. If it can't be replicated in other environments, it will not be of much interest to any funding agent.
- Intuitive
  The WBDC Model must make sense. If it doesn't, it isn't going to be a choice of workers, entrepreneurs, or businesses!
- Robust
  The WBDC must be limited in its application and design only by your own creativity.

Business Incubation Center RBS details

It is clear from the outset that the solution has to link the three environments through a clearinghouse service. It must receive business ideas, projects, process design and improvement and problem solving and make them available to student teams for inclusion in their learning experiences. I'm calling this entity the BIC. What follows is the indented outline of the BIC requirements portion of the complete RBS. I've added the function, sub-function, process, activity, and feature labels for instructional purposes only. In practice those labels would not appear in the RBS. You will also notice that the RBS is not complete because further detail in some of the BIC requirement items are a function of the specific institutional requirements and cannot be part of the template to any degree of specificity. An agile project management approach will be used to manage the design and implementation project and it can identify the missing details through iteration. So the indented outline below is the BIC part of the RBS.

I. Requirement #1: Support the needs of new business formation

A. Function #1.1: Infrastructure
   1. Sub-function #1.1.1: Service level agreement
      a. Feature #1.1.1.1: Services to be offered
      b. Feature #1.1.1.2: Terms & Conditions
      c. Feature #1.1.1.3: Service fee schedule
   2. Sub-function #1.1.2: Membership application
      a. Feature #1.1.2.1: Name of new business
      b. Feature #1.1.2.2: Location
      c. Feature #1.1.2.3: Business Owner
      d. Feature #1.1.2.4: Business products and
         services description
      e. Feature #1.1.2.5: Services needed
      f. Feature #1.1.2.6: Financial situation
      g. Feature #1.1.2.7: Intended launch date
B. Function #1.2: Investigate new business ideas
   1. Sub-function #1.2.1: Business definition
   2. Sub-function #1.2.2: Market definition and size
   3. Sub-function #1.2.3: Market position of new business
   4. Sub-function #1.2.4: Competitor Analysis
   5. Sub-function #1.2.5: SWOT Analysis
      a. Activity #1.2.5.1: Identify strengths, weaknesses, opportunities, and threats
      b. Activity #1.2.5.2: Develop strategies to leverage strengths and opportunities
      c. Activity #1.2.5.3: Develop strategies to minimize impact of weaknesses and threats
C. Function #1.3: Conduct business validation studies
   1. Sub-function #1.3.1: Define business process model
   2. Sub-function #1.3.2: Define and price services
   3. Sub-function #1.3.3: Revenue & expense budget
   4. Sub-function #1.3.4: Cash flow analysis

     5. Sub-function #1.3.5: Establish product/service matrix by competitor

     6. Sub-function #1.3.6: Write business validation report

  D. Function #1.5: Define business plan outline

  E. Function #1.6: Design the new business process process

II. Requirement #2: Support the needs of existing businesses

  A. Function #2.1: Infrastructure support

     1. Sub-function #2.1.1: Service level agreement

     2. Sub-function #2.1.2: Membership application

     3. Sub-function #2.1.3: Service fee schedule

  B. Function #2.2: Project process

     1. Sub-function #2.2.1: Problem solving projects

       a. Process #2.2.1.1: Design problem submission process

       b. Process #2.2.1.2: Establish problem documentation template

     2. Sub-function #2.2.2: Process improvement projects

       a. Process #2.2.2.1: Design business process improvement submission process

       b. Process #2.2.2.2: Establish business process improvement documentation template

     3. Sub-function #2.2.3: Business process design

       a. Process #2.2.3.1: Design business process design process

       b. Process #2.2.3.2: Establish business process-design documentation template

     4. Sub-function #2.2.4: Business process redesign

       a. Process #2.2.4.1: Design business process re-design process

       b. Process #2.2.4.2: Establish business process redesign documentation template

III. Requirement #3: Support the needs of students

    A. Function #3.1: Establish application documentation template

    B. Function #3.2: Establish new business idea documentation template

    C. Function #3.3: Establish Professional Development Process

       1. Sub-function #3.3.1: Establish Professional Development Plan (PDP) Process

       2. Sub-function #3.3.2: Establish PDP Documentation template

IV. Requirement #4: Support the needs of WBDC-owned businesses

    A. Function 4.1: Design WBDC open positions posting process

       1. Sub-function #4.1.1: Design posting document template

       2. Sub-function #4.1.2: Design online posting process

       3. Sub-function #4.1.3: Develop online posting capability

       4. Sub-function #4.1.4: Implement online posting capability

       5. Sub-function #4.1.5: Design position application documentation

    B. Function #4.2: Office support

       1. Sub-function #4.2.1: Allocate private office space for 4 10-person teams

       2. Sub-function #4.2.2: Build-out, furnish and equip each office

       a. Feature #4.2.1: rectangular table

  b. Feature #4.2.2: 10 chairs
  c. Feature #4.2.3: whiteboards on all walls
  d. Feature #4.2.4: internet hook-up
  e. Feature #4.2.5: desktop computer
  f. Feature #4.2.6: telephone service
  g. Feature #4.2.7: flipchart
  h. Feature #4.2.8: key lock on door
 C. Function #4.3: Establish business mentor nomination documentation process
  1. Sub-function #4.3.1: Design application form
  2. Sub-function #4.3.2: Design response to application
 D. Function #4.4: Establish faculty coach nomination documentation process
  3. Sub-function #4.4.1: Design application form
  4. Sub-function #4.4.2: Design response to application

Note how the decomposition structure is used. Function #1.1 uses Sub-function then Feature decomposition. Function #1.2 uses Sub-function then Activity decomposition. Function #2.2 uses Sub-function then Process decomposition. Sub-function #4.2.2 uses Feature decomposition. So the 5-level decomposition structure adapts to any requirement. The ordering of the decomposition levels is fixed but the choice and use of levels is optional.

TRENDS VERSUS SOLUTION REQUIREMENTS

Let's assume that we can put a solution in place that meets all of the requirements stated above. How does it stack up against the trends?

- *Global perspective* - The WBDC Model puts major emphasis on defining and implementing businesses that are sustainable. They have been designed to create barriers to entry for any would be competitors. Technology is a major contributor to that sustainability but technology is not enough. There must also be a creative dimension - one that exploits technology to build those barriers to entry. The BIC, through its curriculum and student team research focuses on creating and defining businesses that are sustainable.

- **Pervasive technology** - In some cases technology is needed just to stay even with the competition. To not exploit technology is a mistake because your competition will. Member businesses will always have access to the latest in business innovation to grow and protect their businesses. The WBDC Curriculum is designed to exploit technology.

- **Systemic business problem-solving** - Problem-solving is as much an art as it is a science. There are several models that can be used but they all require a certain dose of creativity to be effective. The WBDC Curriculum includes a number of such models so that the student teams will be able to choose the best fit model. When a team of students focuses on the same problem but from their own knowledge and experience base they increase the likelihood of finding an acceptable solution.

- **Increased flexibility of educational delivery systems** - The classroom without walls design creates a completely flexible Learning Environment. The instructor is no longer burdened with having to contribute all subject matter content. Their role does include content delivery but their role also includes being a resource to direct students towards sources for additional learning.

- **Increased continuing education** - To be successful in an environment of high change businesses must always be in a learning mode. The WBDC Curriculum is an adaptive curriculum and can respond to meet even the most challenging of business needs.

- **Use of the internet as a learning and marketing tool** -

Because business problems don't come with a suggested solution the solution must be discovered. The solution could be anywhere and the internet is a good first line of attack in that search. At the K-12 level the internet is already being used as a research tool. In the WBDC Curriculum it is an essential ingredient.

- *Partnering with business* - The Learning-Business linkage is an integral part of the WBDC Model. It is a project/problem driven model where learning occurs in response to a specific business need. It replaces the current model where the student team is equipped with a solution (knowledge) out looking for a business application. In the WBDC Model the sequence is reversed and the need for a solution is the motivating force that drives learning.

- *Increased outsourcing* - In the WBDC Model one of the basic premises for business formation is that the business idea be sustainable. That translates into creative uses of technology as well as the use of technology to create barriers to entry for new international competitors. That is the litmus test for all business ideas that are put forth by students or business members.

- *Decreased traditional age students* - Learning is a life-time journey. The six target markets define a diversity of students from those that are inexperienced to those that are very experienced that come from all walks of life and are heading in a variety of directions. That brings a richness of experiences to every student team as well as a solid resource for any businesses that engage student teams to help them with business formation and development.

- *Less manufacturing* - Manufacturing is no longer a growth industry in the U.S. Labor costs in third world countries are significantly less than in the U.S. and have taken away much of the manufacturing business. It is unlikely it will ever return in any scale. There are examples of out-sourcing assembly of a product that from a total cost perspective is less expensive than insourcing the assembly. So any new business idea must account for that and respond accordingly.

- **Sustainability** - There has been a lot of discussion on this characteristic. We all know how important sustainability is to the worker as well as businesses. That is uppermost on our minds as we design and deploy the WBDC Model.

Can you name an operational WBDC Model that possesses all of these characteristics? In all honesty – probably not. I can't. There are certainly models that possess some of these characteristics but not one that includes all of them. Prevailing models are heavily based in the traditional instructor-led classroom. While that may be necessary it is by no means sufficient. Online learning has been considered a disruptive innovation (Christensen, Clayton, 2008, "Disrupting Class: How Disruptive Innovation Will Change the Way the World Learns", New York, NY: McGraw-Hill) and it has shown to improve learning when used in combination with instructor-led delivery or in many cases when used alone. Online courses often tend to automate the lecture approach which doesn't really improve learning. The major benefit of online courses is convenience. Those online courses that use an adaptive learning approach are a notch better but there is still a long road to travel before we can realize the full impact of online courses on learning. Even where online learning is effective it doesn't provide a complete WBDC environment. Other models include projects, internships and work/study components but few of these programs deliver the promised integration between business and academe. Many work to learn programs tend to be unstructured with little or no actionable results. They appear more as add-ons to a program because that is the cool thing to do. What is needed is a creative replacement of our Industrial Age education and training delivery system. The proposed WBDC Model coupled with the concept of a BIC is a disruptive innovation that stands unique among WBDC Models. I predict that it will usher in a new and invigorating approach to career and professional development for every worker regardless of their station in life and their career and professional development goals.

What is needed is a creative replacement of our Industrial Age education and training delivery system.

While I can appreciate the complexity involved and understand the interdependencies that exist among the three environments and the BIC I believe I have formulated a comprehensive and adaptive solution that will work. A major feature of the WBDC Model is that it is scalable, adaptive and robust. The program depends on the active involvement of local businesses. As the program is launched it will work even if the number of participating businesses is small. There will be plenty of project-based and problem-based challenges for the students. As the number of students in the program increases the Model actually broadens and deepens. The number of program-owned non-profit businesses will grow and the Incubation Center will provide more work options for the students. While most business development will be focused on leveraging technology there are several possibilities for other types of businesses.

PUTTING IT ALL TOGETHER

To be effective and have some longevity the initial solution must have all of the requirements discussed above. Its effectiveness will come from being adaptive to the special needs of workers, entrepreneurs and business owners. Its longevity will come from accommodating the current trends and hopefully unnamed trends that will make themselves known.

CHAPTER **3**

The Solution As I See It

A viable solution to workforce and business development must be comprehensive and adaptive. The solution will have many interdependent parts and will not be easily implemented but it is a solution and, unless someone has a better idea, it is the best opportunity we have for implementing an acceptable solution. It will require stepping outside our collective comfort zone and a radical rethinking of how we educate our students, train our workers, and support small business formation and development.

As far as comprehensive workforce training and business development are concerned I believe that the current delivery systems must change or risk being dismissed as irrelevant. The monolithic delivery models that are so commonly used in delivering education and training programs are simply not cost or time effective. Except in simple training situations I don't think they have ever been effective. Attendees at training sessions tend to evaluate their experiences there more from an entertainment perspective than from a learning perspective. I know several trainers who get rave reviews for their workshops and training but in truth the content is weak and the entertainment value is high. Certainly entertainment value is necessary but it is by no means sufficient.

The monolithic delivery models that are so commonly used in delivering education and training programs are simply not cost or time effective.

Regardless of your feelings regarding our government's attempts to restore our economy one thing is certain - every state desperately needs to train entry level workers and retrain career changers. And this need reaches over to small business formation and development. The gauntlet has been thrown and the education sector needs to take a serious look at itself and be creative in its approach to vocational and professional education. Let's not let our thinking be shackled by the old models from the Industrial Age but rather begin thinking outside the box at the possibilities.

The spirit of innovation is not lost in America, it is just hibernating and it is time to wake it up.

Small businesses have been particularly affected by the downturn in the economy. The federal government recognizes small business formation as the engine that drives our economy and yet our elected officials have not stepped up to the bar and put programs in place to get that engine going again. The WBDC hopes to rekindle that energy and provide the guidance and support needed to put small business development back in the mainstream.

THE VISION OF THE SOLUTION

The WBDC Solution is above all else a lifelong career and

professional development service designed to provide the one-stop resource for individuals seeking guidance and support for their lifetime career and professional journey and the one stop resource for new and developing businesses needing ongoing support from defining to planning to launching and finally to reaching maturity.

... the one stop resource for individuals seeking guidance and support for their lifetime career and professional development journey and the one stop resource for new and developing businesses needing ongoing support from defining to planning to launching to reaching maturity.

The WBDC Solution provides a lifetime of job support to individuals. That includes learning, discovery and application through actual business experience. Through the WBDC Model a person can develop and maintain a lifelong career plan and draw upon WBDC resources to maintain and achieve that plan.

The WBDC CareerAgent (Figure 3.1) is one example of a career planning infrastructure that will be a major component of the WBDC Solution. The WBDC version will include a Professional Development Plan (PDP) which is a planning and deployment tool that the individual can use over their entire career. WBDC CareerAgent will help the person answer the following questions:

- Where are you?
- Where do you want to go?
- How will you get there?
- How are you doing?

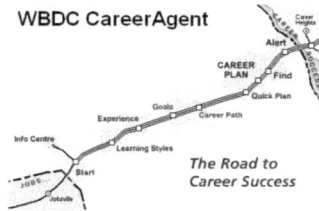

Figure 3.1: WBDC CareerAgent

WBDC CareerAgent has a history dating from 1994. I proposed the application in 1994. I managed the client-funded
project and in 1997 my team implemented a thin-client internet application. In less than 6 months CareerAgent enrolled over
18,000 IT professionals looking for career and professional development planning and support. A need to redirect resources to
their core businesses resulted in the client selling CareerAgent.
CareerAgent was taken off the market, dissembled and repur-
posed as an online candidate self-evaluation system. I own the
intellectual property of CareerAgent as shown in Figure 3.1 and
would support bringing it back as a robust career and professional planning system for any organization wishing to implement my
proposed WBDC Model. Designing, developing and deploying
WBDC CareerAgent would be a challenging development project
for a student team.

The Short-term Goal of the WBDC Solution

By the end of the third year of operation the WBDC will be
a self-supporting program serving the professional and business
development needs of the six target markets defined in Chapter 1.
These six groups are very different with very different levels of maturity, experiences, needs and motivations. The WBDC Solution will
be sufficiently robust to reach out to all of them through its unique
project-based team-centric learning and development model. The

WBDC Solution architecture is robust and can be adapted to any discipline and any type of business (such as family-owned product and service businesses).

An Overview of the WBDC Solution

In this section I present the WBDC Solution that I have not seen elsewhere. I have seen parts of the model in application and have some idea of what works and what doesn't work. From my own direct involvement in career and professional development I have crafted the WBDC Solution described below. I have not yet had an opportunity to discuss this model with any credentialed education researchers and I would hope to have an opportunity to participate in such discussions because I know that will lead to further improvements in the model.

The WBDC Solution fully integrates academe, business and business development to create a team-centric training and retraining experience for students and workers. Each team is typically comprised of 5-6 students with common career and professional development business interests. Team strength and effectiveness follows from the integration of the six target markets into a single team. An individual remains on a team until they complete their program. Their team affiliation can change in order to continue accommodating their learning needs. The team will be supported by a faculty advisor and mentor from the business community. A team may be attached to a single WBDC-owned non-profit business or new business idea. More than one team may be attached to the same business. Team membership changes as students complete their programs and new students enter the program. New students to the program will interview to join a business just like they will someday interview for a real job! Businesses operate in a team-centric mode just as the WBDC Solution is a team-centric structure. The BIC is as much a microcosm of the real world as it can possibly be! It provides as rich a source of learning and discovery opportunities as is possible.

The WBDC Landscape depicts a unique and innovative learning and training environment. I am not aware of a similar program in the US. It is based on established adult learning theory and is designed to incorporate all four adult learning styles. The four adult learning styles are analogous to the following situations for someone who is learning to swim:

- **Observe:** I'll just sit on the side and watch how people swim.
- **Experiment:** I'll get in the pool but I need someone at my side to keep me from sinking.
- **Reflect:** First I'd like to read a book on the buoyancy of solids in a liquid medium.
- **Experience:** Just throw me in and I'll figure it out on the way down.

These are all very different styles of learning. Over the years I learned through experience that a training program will not be effective if it does not offer the student all four learning opportunities. The WBDC Curriculum that I envision will do just that.

There are three environments and three linkages that describe the WBDC Landscape which is shown in Figure 3.2 below. The most important thing to note about this landscape is that the BIC is part of every linkage. In fact, it is the bridge that provides the application link between each pair of environments. The effective functioning of those linkages is essential to the successful operation of the WBDC.

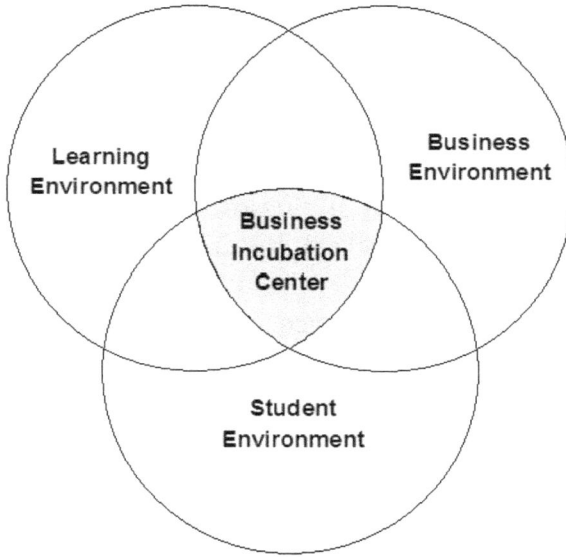

Figure 3.2: The WBDC Landscape

The four components of the WBDC landscape

The WBDC landscape defines the environment within which programs for students, workers, entrepreneurs and business owners will be built and thrive. The result will be to bring all four of these populations back into the mainstream of our economy. There are four major components to the WBDC landscape as defined below.

- Learning Environment

In the WBDC Landscape the classroom is not what you might think it is. The WBDC Solution is designed using the concept of a "classroom without walls". In the WBDC Model the classroom is any place where learning needs to occur, can occur and is planned to occur. That would include the traditional classroom as the focal point but it is a much richer

Learning Environment than that. Under the guidance and advice of their faculty and business advisors the team will have identified a learning objective and it is their responsibility to develop and gain faculty approval of their Learning Contract to acquire it. They will have to go wherever they need to go and get whatever they need to get to acquire that learning. It may be the traditional classroom but it might also be a local business, the internet, courses at another university, a museum, an expert located anywhere in the world or even attendance at a conference or professional society meeting. Chapter 4 is a more detailed discussion of the Learning Environment.

- Business Environment

The Business Environment contains two separate environments: the Entrepreneur Environment and the Business Owner Environment. Each has a different perspective and motivation for using the WBDC. The entrepreneur wants to define a sustainable and profitable business but needs help validating the business idea, forming and launching the business. The business owner wants to stay in business but needs business process help, problem-solving support, and customized research.

- Student Environment

Students and workers enter the program with some idea of where they want to go. (When the worker enters the WBDC Program they become a student and will be referred to as a student throughout the book.) The student may not have the specifics but at least they know whether their future lies in growth in their existing position, continue working for others or in starting and managing their own small business. They look to the WBDC Program to help them fill in the blanks and get their plan for career and professional development in place and underway. As they work on their plan they will turn to the WBDC for support and guidance.

- BIC

The WBDC BIC is the heart and soul of the WBDC. I am not aware of a similar entity in operation anywhere in the U.S. It does offer the support that you ordinarily associate with an incubation center but it goes much further.

The WBDC BIC is really the meeting place for student teams, teaching faculty, entrepreneurs and business owners. Student teams are looking for:

» preparation for first-time entry or re-entry into the world of work
» practical application of learning on real projects
» the use of real projects to drive and motivate their learning program
» advice and coaching from member business owners

Teaching faculty members are looking for:

» participation from businesses in the learning process
» projects from entrepreneurs and business owners to drive the learning process
» support identifying internet resources

Entrepreneurs are looking for:

» an unbiased review of the feasibility of their idea
» an evaluation of the proposed business
» help in getting the new business defined, planned and launched

Business owners are looking for:

» help in researching a new line of business
» process design and improvement assistance
» leveraging technology to protect and expand market share

If you take a step back and look at the BIC you will see that it is an ever-changing microcosm of the real world of business.

The BIC is the heart of the WBDC Model and it provides

for the integration of the three environments using the BIC as the bridge. It is the bridge that allows the three linkages to function as a living organism and produce the expected results.

At any point in time the BIC will contain a wide range of projects that will be used by student teams to drive their learning experience and simultaneously meet the needs of the entrepreneurs and business owners who submitted them. The BIC project portfolio will include:

> » Student and team-proposed new business ideas
> » Entrepreneur-proposed new business ideas
> » New product/service development projects
> » Team-based non-profit businesses owned by the WBDC
> » Business owner member's process design/improvement projects
> » Entrepreneur and business owner member's problem solving projects
> » Research-based contract work for member businesses

The number of potential technology-based business ventures in the BIC is limited only by the creativity of the WBDC students, business members and other participants. Three ideas that come to mind immediately are:

> » The automation of health care delivery systems, i.e., online meal ordering options, paperless operations, consolidated patient data collection and distribution
> » The design and development of a web-based career planning system
> » Applications of animated graphic solutions to specific business process design and decision making models, i.e., dispatching and delivery, order entry and order fulfillment

The BIC will be a microcosm of the real business world. It

will house a number of non-profit businesses at all stages of the business life cycle. These businesses will be managed by student teams under the advisement and mentoring of a faculty member and a manager from a member business. Teams will apply the principles and concepts delivered in class to actual business situations. Conversely, business projects and problems will arise that can be used in courses to drive and motivate the process of learning and discovery.

The BIC will also be the depository of business process design and improvement projects and unsolved business problems. These are the stimuli for learning and discovery that will drive course content. The projects and problems will be a community asset – available to any team for inclusion in their course(s). Non-disclosure and non-compete agreements will be used at the request of the entrepreneur or business member.

The WBDC Model also encompasses business formation of any kind. It is the "skunk works" of the WBDC Model. While a technology base is the common thread of many businesses it is not the only common thread. For example, there has also been interest expressed in a family-business focus and that can certainly be supported by the WBDC Model.

I see the BIC as the focal point of the WBDC Model. It is a dynamic source of applications for every course in the WBDC Curriculum.

The BIC is a living laboratory for learning and discovery. The needs of the student staffed business will drive the projects and problems studied in the classroom. Conversely, the concepts learned in the classroom will be applied to the businesses in the BIC. In effect the BIC businesses become living laboratories for learning, experimentation and discovery for all the students. The BIC functions as the clearinghouse for business projects and problems. It is these projects and problems that drive the learning and discovery processes of the WBDC curriculum.

The three linkages of the WBDC components

The WBDC Model contains the BIC along with the three environments described above. They form a fully integrated and interdependent set. Furthermore, these three linkages form a necessary and sufficient set for a comprehensive solution to be delivered by the WBDC Model. Note that all three linkages are two-way linkages. Each linkage can include the BIC as the actual link between the two Environments.

- Learning to Student Linkage

The student depends on the Learning Environment to provide the programs and curriculum they will need to get into the workforce, stay in the workforce and grow in the workforce. All that the student expects is to not have their time wasted, to get good advice and the program that they need regardless of their avocation. Chapter 4 explores this linkage in more detail.

- Learning to Business Linkage

This linkage provides yet another venue for learning and discovery by student teams. It is a two-way linkage. Business executives will be invited to attend and speak in the classroom on topics of interest to the subject matter being studied or observe and critique team presentations. Businesses will reciprocate by inviting teams and students to visit them in the workplace. The teams might be making presentations during these visits or observing certain business meetings or activities. The students may be gathering data and information for the team for their WBDC Project. Through this linkage business and academe will be fully integrated into the process of learning and discovery. Chapter 5 explores this linkage in more detail.

Business to Student Linkage

The role of the BIC is most obvious in this linkage. Student teams look to the BIC for projects they can use as the driving force in the formation of their learning programs. That is the fundamental idea behind project-driven learning. Business owners look to the BIC as the depository for problems, process design/improvement and other projects for student teams to work on. So the BIC functions as a clearinghouse for student teams and business owners to get their needs met.

It's a win-win linkage. Chapter 6 explores this linkage in more detail.

WBDC Process Flow

Figure 3.3 illustrates the WBDC process flow.

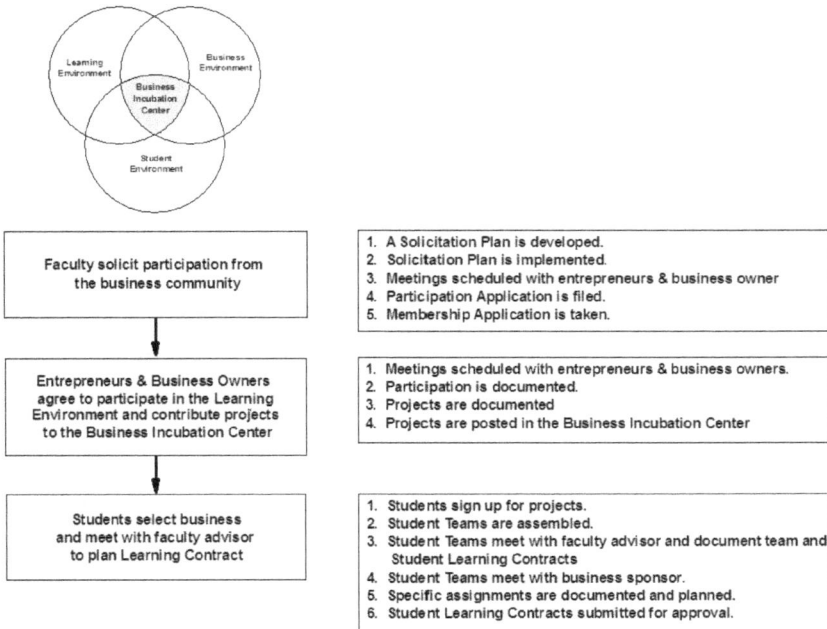

Figure 3.3: WBDC Process Flow Diagram

The Learning Environment to Business Environment linkage marks the start of the WBDC Process. This part of the process operates continuously. The faculty and administrators need to take the initiative and establish a continuous infusion of direct entrepreneurial and business owner experiences into the classroom as well as a continuous supply of projects into the BIC.

Meetings are scheduled with each entrepreneur and business owner who agrees to participate in the WBDC. For the experiences and presentations they wish to share in the classroom a document describing that participation will be needed. That document becomes a formal part of the course(s) to which it is attached. For the projects they wish to contribute to the BIC a detailed description of the project is required and is posted in the BIC for student review and consideration. Company names are not disclosed at this time. In order to protect the entrepreneur proprietary information is not disclosed either. The document becomes the basis for aligning students with projects and incorporating the projects into the student's Learning Contract. The Learning Contract is discussed in Chapter 4.

The posted projects are input to the student planning their learning program. Based on their career interests they will select one or more projects. The projects they select might be aligned with a particular business area of interest or a business function they need to understand. The faculty will form student teams based on input from the students. These teams will meet with the business sponsor responsible for submitting the project to the BIC. The projects are integrated by the student into their Learning Contract and submitted to their faculty advisor for final approval prior to implementation.

WBDC Model Deployment Strategy

The transition from whatever model you are now using to this WBDC Model is significant. If I were to consider implementing it, I would spend considerable workshop time with the key participants assuring they understand the scope of the undertaking and the fact that it is an adaptive effort. Here is a place where an honest and open project scoping and requirements elicitation exercise is an essential ingredient for success. Figure 3.4 is my recommendation for a deployment strategy. It is only a template that a specific institution will customize to meet their exact specifications. Let's take a brief look at each activity in Figure 3.4. A more detailed discussion is provided in Chapter 7. A complete Guide for designing and implementing the WBDC is available from EII Publications. Those who purchase the Guide will receive an extensive file of electronic templates. Email me at rkw@eiicorp.com for details on acquiring the Guide.

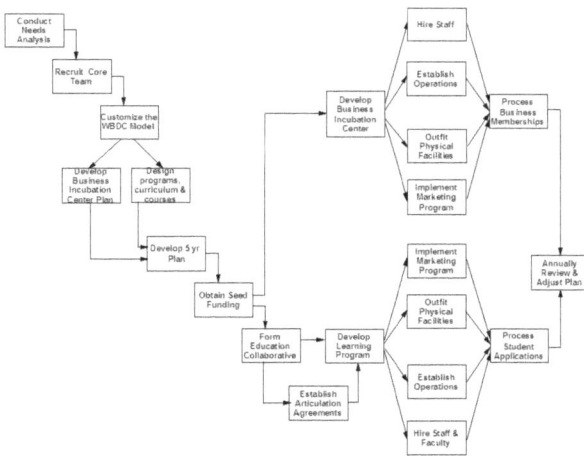

Figure 3.4: A WBDC Model deployment strategy template

Conduct Needs Analysis

The Needs Analysis can be conducted by the institution or with the assistance of EII consultants. Typical needs analyses require a two day onsite workshop. The workshop agenda consists of answering six questions:

- What is your market situation?

- What do you need to do?
- What can you do?
- How will you do it?
- How will you know you did it?
- How well did you do?

This Needs Analysis is sufficiently robust so that it can be used for new business formation as well as business development for existing businesses.

Recruit Core Team

With some direction provided by the Needs Analysis a Core Team is formed. The Core Team consists of representatives from the major administrative and academic departments that will provide input to or adapt the output of the WBDC. These representatives will remain on the project team for the entire project. Other team members will be added as needed. The Core Team should be directed by the director-designate of the WBDC.

Customize the WBDC Model

The WBDC Model described in this book and further described in the Guide is a template. To be useful to the adopting institution it must be customized. There are two distinct parts to that customization: develop the BIC Plan and design the programs, curriculum and courses.

As part of this customization a draft document describing the mission, vision, and objectives of the WBDC should be prepared. The expected products and services that will be provided by the WBDC should also be decided and briefly described. An organization chart should be included if appropriate.

- Develop the BIC Plan

I like to think of the BIC as the engine that drives the WBDC. It is a comprehensive microcosm of the real world of business. Developing the plan to implement it is a challenging project.

The danger is in developing a steak solution on a baloney budget. My recommendation is to base its implementation on a phased-plan. The phases might look something like this:

Phase 1: Define the BIC services

Phase 2: Construct and furnish facilities

Phase 3: Prioritize the target markets

Phase 4: Develop and execute the marketing plan for the top priority target market

Phase 5: Repeat Phase 4 for the next target market

Phase 6: Assess performance measures and revise BIC as appropriate

- Design WBDC programs, curriculum and courses

The WBDC will offer a limited core curriculum in career and professional development planning and entrepreneurship. A Certificate in Business Formation and Development will be its only program. Other minor, certificate and degree programs will be done within articulation agreements with the home institution or local institutions.

Develop 5 Year Plan

The 5 Year Plan is a detailed document that describes how the WBDC Program will be designed and implemented. It includes revenue and expense analysis and a cash flow analysis.

Obtain Seed Funding

Using the 5 Year Plan a program to raise the needed seed funding is launched. The template assumes $2M will be needed. That number will be adjusted by the Core Team depending on the institutional situation. The seed funding should be sufficient to bring the WBDC to a profit position by the end of the third year of operation.

Form Education Collaborative

The WBDC Program will most likely be built on the assumption that a collaborative of local educational institutions will be required in order to implement the curriculum.

Establish Articulation Agreements

Articulation agreements are commonly used between institutions that will accept course credit for courses taken at one institution as satisfying program and degree requirements of the other institution.

Develop BIC

This is a major part of the WBDC Program. The plan would have been described in the customized version of the WBDC Model completed earlier. Developing the BIC consists of the following activities:

- Implement Marketing Program
- Outfit Physical Facilities
- Establish Operations
- Hire Staff

Develop Learning Program

The WBDC Curriculum would have been defined and course descriptions written as part of the customized version of the WBDC Model completed earlier. Developing the Learning Program consists of the following activities:

- Implement Marketing Program
- Outfit Physical Facilities
- Establish Operations
- Hire Faculty & Staff

Process Business Membership Applications

Business members will be either entrepreneurial ventures being investigated for feasibility and formation or they will be existing businesses looking for services from the WBDC. Membership applications will be sought from the local business community.

Process Student Applications

An aggressive marketing program will have been conducted and applications being submitted as a result.

Annually Review & Adjust the Plan

The implementation of the WBDC is a continuous process. It will never end. At periodic intervals (probably annually) the performance against a defined set of quantitative metrics will be assessed and adjustments made as needed.

PUTTING IT ALL TOGETHER

As far as I know the WBDC Model that I have described is unique. I envision it as a dynamic and robust program. In defining its contents and delivery process we, as educators and trainers, will be challenged to constantly re-invent ourselves and are limited only by our own creativity. Because the WBDC Model is based on a team-centric and project-based learning model it will automatically be aligned to the needs of business and produce graduates who have demonstrated through actual WBDC-based experiences that they can fill those needs. Having had this experience as part of their education and training is a powerful credential and should serve the worker as they enter the world of work.

...we, as educators and trainers, will be challenged to constantly re-invent ourselves and are limited only by our own creativity.

But the WBDC Model goes even further. It is designed to support the worker over their entire career. Things will change and technologies once thought to be necessary will be replaced by even more powerful technologies, new opportunities will arise and the cycle will repeat itself over and over again. Career and professional development is a lifelong journey. The WBDC Model will also adapt and be there for lifelong support of the worker.

Features of the WBDC Model

- WBDC Model is a unique and innovative idea that will allow us to rekindle the creative energies that built this great nation. To my knowledge there is no equal anywhere in the world.
- WBDC Model offers a completely reinvented education/training delivery model
- WBDC Model leverages the excitement of projects and problems to enhance learning
- Through the BIC the WBDC Model provides a living business laboratory for learning, discovery and rekindling of that "spirit of innovation" that President Obama has frequently spoken about. It is a benefit to all.
- WBDC Students are organized into teams coached by a faculty and business person
- WBDC Model uses a team-centric learning model
- WBDC-owned not-for-profit businesses are team operated and professionally supervised
- WBDC Model offers a fully integrated state certified curriculum at the high school levels
- WBDC Model provides a fully supported environment for new business development
- WBDC Curriculum can be integrated into existing high school and university programs
- WBDC Students learn about the world of work in a safe mode
- WBDC Model is scalable and can be replicated in other cities, states and disciplines.

Benefits

The major benefit of the WBDC Model is that it creates safe harbor for all workers, entrepreneurs and business owners. While their needs are quite diverse and continuously changing the WBDC is robust and adaptive and can accommodate all three markets.

A secondary and equally important benefit is that it establishes the disruptive innovation foundation for the education and training communities to meet the changing needs of their changing markets.

CHAPTER **4**

Learning Environment

The Learning Environment provides for the staging, definition, approval, and execution level support of the Student Learning Contract (SLC). There are two linkages that are utilized to make this happen: the first is the Learning Environment to Business Environment linkage, the second is the Learning Environment to Student Environment linkage. The first linkage defines and documents projects and posts them in the BIC. The second linkage draws projects from the BIC and uses them to define the project-driven SLC.

This is the first of three chapters that are needed to fully describe the WBDC Model given in Figure 3.2 and reproduced here with more detail as Figures 4.1 and 4.2. In this chapter the Learning Environment is defined and discussed in detail. Then the two linkages it establishes are defined and discussed in detail.

THE LEARNING ENVIRONMENT

The best way to envision the Learning Environment is to think of it as a "classroom without walls." The term "A classroom without walls" has been used before and so it is not unique to the WBDC but its definition and operational activities are unique. The first and most important responsibility of the Learning Environment is to establish a participatory relationship with the Business Environment. The Learning Environment must draw the Business Environment into the learning experience. Part of this is the direct involvement of businesses in the "classroom." Classroom here is interpreted in the broadest context - a classroom without walls. The second responsibility is to establish a rich depository of projects submitted by local businesses and entrepreneurs. That depository provides the lifeline to the project-driven curriculum.

The Learning Environment also encompasses a collaborative formed of participating high schools, community colleges and universities. They will all contribute courses and programs to varying degrees of detail. Of prime importance will be the articulation agreements that describe the degree granting relationship between pairs of participating institutions. A student from any of the participating institutions can take advantage of the WBDC Curriculum to enhance and enrich their program. Articulation agreements have been used for several years and are an accepted way to complete program requirements.

WBDC Projects

The WBDC Curriculum will be a project-based curriculum that is delivered in a team format. This is a big step outside the comfort zone of many faculty and their administrators. It is in fact a disruptive innovation in post-secondary educational delivery systems. It is not to be treated lightly but rather approached with great consideration of the impact it will have on an institution and based on a healthy dose of due diligence. The BIC, which is part of the Learning Environment, accepts and maintains the approved projects.

These projects can come from students, faculty and the business community. Students will propose new businesses ideas as part of their application process, faculty will have acquired projects from their colleagues, and businesses under the guidance of a faculty member will submit projects to meet their business needs.

Projects may extend for more than a semester and may be used by one or more student teams in their SLC. In the discussion below the term project is used in its most general sense and includes one or more of the following:

- Feasibility studies
- Market analyses
- Competitor analyses
- SWOT analyses
- Business plan documentation
- New process design
- Process improvement
- Problem solving
- Customized research

There are specific criteria that an undertaking must meet in order to qualify as a project suitable for inclusion in the BIC. Those criteria have been previously defined (Thomas, John W., 2000, "A Review of Research on Project-Based Learning," The Autodesk Foundation) and are adapted to the WBDC Model below:

WBDC projects are central, not peripheral to the curriculum.

To put it another way, a WBDC course is not a course out looking for a project rather it is a project out looking for a course or courses to support execution of the project. The project provides the vehicle for learning one or more disciplines. WBDC Projects are not examples of the application of concepts or principles learned in class. Finding such projects that qualify for the WBDC Model is a challenge. The challenge is somewhat nullified by the BIC and the meaningful involvement of the business community.

WBDC projects are focused on questions or problems that "drive" students to encounter (and struggle with) the central concepts and principles of a discipline.

Usually this is the result of a poorly defined or incomplete question or a question that involves two or more disciplines. Unsolved business problems are often good sources for these projects. They are unsolved because of their complexity and exceptions to the norm.

WBDC projects involve students in a constructive investigation.

The investigation involves inquiry, knowledge building and resolution. In other words, it involves the construction of knowledge. If it only involves the application of already-learned knowledge, it does not qualify as a WBDC Project. At the outset it is not possible to chart the direction the investigation will take. The answer to one question will often give rise to several other questions that require answers. The project-driven curriculum is a rich source of motivated and practical learning.

WBDC projects are student-driven to a significant degree.

WBDC Projects are adaptive projects. That is, the outcome is not pre-determined as in the case of laboratory exercises. A WBDC Project often gives rise to a need to craft creative approaches to projects and problems they encounter. The approach may require more than one discipline. Therein lay the motivating factors. Student involvement will certainly be heightened by their active participation in the process of learning and discovery because they see an immediate application.

WBDC projects are realistic, not school-like.

For all intent and purposes a WBDC Project is a real project that could be implemented. It has all of the characteristics and feel of reality. Again the BIC will be a great source for such projects. While the instructor may be able to generate suitable projects through their own efforts it is more likely that they will depend on the BIC for most of their ideas. I expect that projects taken from the BIC will have a higher value to the student teams than will a project developed by their instructor. I also expect that the instructor will expand the scope of a project taken from the BIC in order to expand the knowledge content of the project.

The local business community is a resource to the BIC businesses. Their advice and opinions will be actively sought. They have "been there and done that" and will be invaluable to the teams by bringing the real world into the BIC embryonic business and business ideas. Every business in the BIC will have a mentor from the business community and a faculty sponsor.

The BIC is also a resource for the business community. Here is the place where new business ideas can be tested in a skunk works setting. Student teams can be commissioned to research new business ideas, new/revised business processes and other feasible ventures in a low cost and no risk setting for any of its business partners. Businesses can use the BIC as a permanent demo site and a place to hold training for their employees and presentations to their staff and customers.

WBDC Courses

There will be a selection of core courses in all programs (Appendix B). These will be offered out of the WBDC. Other formal courses will be drawn from existing courses offered by the participating institutions. In collaboration with the participating institutions that have articulation agreements in place with the WBDC, students will file and gain approval of degree or certificate programs according to their home institution's graduation requirements. The content of the SLC will align with these requirements. Due to the adaptive nature of the programs there will be independent study and project-based learning components but still compliant with degree requirements.

Depending on the needs of the student a selection of core courses is chosen in consultation with a faculty advisor. The core courses that I propose for the WBDC Curriculum are:

- General Required Core Courses (11 cr)
  - » Career and Professional Development Planning Workshop (1 cr)
  - » Introduction to Entrepreneurship (1 cr)
  - » Fundamentals of Effective Project Management (2 cr)
  - » Fundamentals of Business Analysis (2cr)
  - » Creative Problem-Solving Methods (2 cr)
  - » Written and Verbal Communications Skills (2 cr)
  - » Effective Team Formation and Building (1 cr)
- Entrepreneurship Required Core Courses (6 cr)
  - » Fundamentals of Creativity and Innovation (1 cr)
  - » Marketing Research (2 cr)
  - » How to Write a Business Plan (2 cr)
  - » Financing the New Venture (1 cr)

These 11 core courses total 17 credit hours and could form an Entrepreneurship Certificate Program or a minor within a BA or BS program in any professional degree program and most disciplines. By adding a few electives a BA or BS in Entrepreneurship can be defined. Other formal courses will be drawn from existing courses offered by the participating institutions. In collaboration with the participating institutions students will file and gain approval of degree or certificate programs according to the program requirements of their home institution. Due to the adaptive nature of the programs there will be a heavy reliance on project-based and problem-based courses. The content of these courses will be adaptive but still maintain compliance to their home institution's program requirements. Because of the adaptive nature of the WBDC Curriculum these will be defined at the appropriate time in each student's program.

I don't want you to get the impression that the WBDC Curriculum is a free for all curriculum. It is adaptive because it uses project and problem-based learning models. Courses derive their content and structure from the needs of the student and the projects contributed by businesses and others but there is always an underlying learning requirement defined by their program requirements. So there is a standard that defines what disciplines, concepts, principles and theories must be included in the certificate or degree program but not how that is included in the program. The how is the adaptive part of the program guided by the projects in the BIC. This changes the rigid requirements of taking a defined collection of cataloged courses to a program designed to meet specific learning objectives. This is a major change from the offerings of traditional institutions but it offers a closer alignment to student needs than current programs. Obviously a faculty advisor has much more authority than current models that are built around a static set of courses. They are given the authority to approve programs of study that they have decided meet institutional requirements. This differs from the traditional institution that requires the approval of a curriculum committee. Those models are obsolete and don't meet today's needs.

There are two learning models to discuss: project-driven or course-driven.

Project-driven learning model

This is the preferred and most effective of the two learning models. The need to design a business process is a project and already exists in the BIC having been submitted by one of the member businesses or would have come about in the investigation of a student business idea. A student team will have aligned with the business process design need and turned to the curriculum to help them learn how to design and document the requested business process. To the extent possible the project-driven learning model will be the model of choice.

Course-driven learning model

In the absence of an existing process design request from a member business the students will have taken the course work and now must find an application of their new-found knowledge. This is the old model where the project follows the knowledge acquisition. This approach is far less motivating to the student teams because they do not see an immediate application while they are charged to acquire the knowledge. Applications are often contrived to provide that motivation but nothing substitutes for the preferred model.

A fundamental premise of the WBDC is that learning is not effective if it doesn't include application. It can be application-driven, learning-driven, or concurrent.

- *Application-driven* - Application-driven learning (a.k.a. project-driven or problem-driven) occurs when a team is presented with a need for knowledge to complete a specific business activity. The team knows it doesn't know how to complete the activity but it knows where to go to find out.
- *Learning-driven* - The student team has identified a need to learn something and turns to the BIC to look for a solution approach.
- *Concurrent* - This approach to learning is a more directive approach. The instructor will pose a problem, process design, or process improvement activity and some approaches that might be useful. The student team is then left to its own initiative to resolve the situation.

THE LEARNING ENVIRONMENT TO BUSINESS ENVIRONMENT LINKAGE

A relationship between faculty and businesspersons is not unique to the WBDC but the relationship defined in the WDBC Model is unique. Current learning models include internships and businessperson involvement in the classroom as part of their business relationships. These are more of add-ons to the learning experience than they are essential components that enrich the learning experience and even make it possible. The WBDC goes much further than these current models as you will see. Figure 4.1 illustrates the linkage.

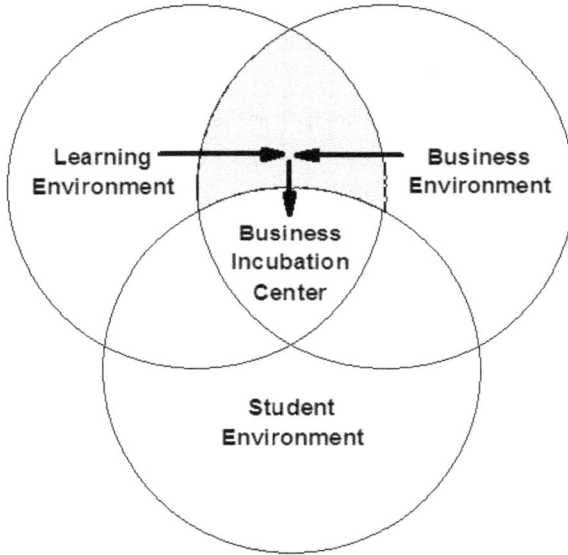

Figure 4.1: Learning Environment to Business Environment Linkage

The Learning Environment to Business Environment linkage is used to establish the foundation and resource availability from the business community for the WBDC learning experience. There are five types of deliverables that flow from this linkage:

- The identification and documentation of classroom experiences that involve the participation of the entrepreneur and business owner in the classroom
- The identification and documentation of projects that are then submitted for posting in the BIC. These are carefully worded to protect the identity of the sponsoring business.
- This linkage is usually from the Entrepreneur Environment to the Learning Environment. Entrepreneurs are invited to

appear in class to present their thinking on business formation. Of particular interest will be discussions on creativity
and innovation as precursors to business formation.

- Small Business Development Centers (SBDC) will often be
invited to send a representative to the classroom to discuss
the Small Business Administration (SBA) and the role of the
SBDC in business formation. Of particular interest will be
discussions on mitigating the risks associated with business
failure.

- Informational interviews which can be preparatory of further research and study by the student team relevant to a
project they are working on or they might simply be career
exploration.

Involving the entrepreneur and business owner in the classroom

Faculty and administrators are responsible for initiating
relationships with the business community that will produce commitments on the part of the entrepreneur and business owner to
participate in the Learning Environment. This participation can
take any one of the following forms:

- Class presentation by the entrepreneur or business owner
on any of the following topics: new business idea, business
change to protect markets, other news worthy events, and
topics requested by the students.

- Panel discussion on contemporary business issues such as
innovation and creativity, leveraging technology for sustainability, leveraging technology for business formation
and development, as well as others.

- Reviewing and evaluating student presentations

- Teaching the application of a business process, tool, or
template to their business

These will be documented and linked to the appropriate
courses.

This linkage is purely for knowledge acquisition. That happens through the participation of the learner in the Business Environment and vice versa. Any of the following events can deliver a learning experience.

Business Environment events

These learning experiences will mostly involve the student team being invited to participate in business meetings. Their participation will be totally passive. They are not expected to provide any comments or input in the meeting. Their role during the meeting is to observe and take personal notes if they wish. Following the meeting they may have a debriefing session with their business sponsor to discuss what took place and answer any questions. This would be an opportunity for the students to provide input.

Learning Environment events

These learning experiences involve a business person being invited to actively participate in a learning event. These will usually take place in a classroom, lecture hall or small conference room. Some typical events are:

Critique a student team presentation

Discuss applications of a tool, template or process

Make a presentation on a topic suggested by the student team

Submitting projects to the BIC

Faculty and administrators are responsible for initiating relationships with the business community that will produce projects submitted by the entrepreneur and the business owner to the BIC under the advice and direction of the faculty. The projects are documented and submitted to the BIC for use in WBDC courses. These projects are classified into one of the following types: new business formation projects (submitted by entrepreneurs) or business development projects (submitted by business owners).

The WBDC faculty and administrators are expected to actively pursue businesses to have them contribute projects for inclusion in student learning programs. These projects can be related to business formation or business development. Business formation projects include feasibility studies, business validation studies, SWOT Analyses, business process design, business planning and specific research projects. Business development projects include process design, process improvement, problem solving and specific research projects.

The local business community is a resource to the BIC businesses. Their advice and opinions will be actively sought. They have "been there and done that" and will be invaluable to the teams by bringing the real world into the BIC embryonic business and business ideas. Every business in the BIC will have a mentor from the business community and a faculty sponsor.

The BIC is also a resource for the business community. Here is the place where new business ideas can be tested in a skunk works setting. Student teams can be commissioned to research new business ideas, new/revised business processes and other feasible ventures in a low cost and no risk setting for any of its business partners. Businesses can use the BIC as a permanent demo site and a place to hold training for their employees and presentations to their staff and customers.

New business formation projects

There are a variety of new business formation projects that will create excellent project-driven learning experiences for student teams. They include:

- **Feasibility studies** - The basic question here is does the business idea make sense
- **Market analysis** - What is the size of the market and what similar products and services are available to that market? Are some of the needs of that market not met with the current products and services?

- **Competitor analysis** - Knowing your competitors and how they provide product and service to the market is critical to the decision to continue to investigate the business idea. The question to be answered here is "is there room for a new entry in the market?"
- **SWOT analysis** - This service provides an objective view of the business idea. The SWOT analysis will include a strategy for mitigating the weaknesses and threats and taking advantage of the strengths and opportunities as part of configuring the business for success.
- **Product, promotion, pricing, positioning research** - At this point in the business formation process the business has been found to be feasible and worth taking to the next level of description.
- **Financing a new venture** - There are several options for financing a new business venture. SBDCs are a good start for SBA loan. Local banks and venture capitalists are another. There are special programs for women and minority owned businesses. Under the guidance of the entrepreneur a student team will research the possibilities and assist with application preparation.
- **Cash flow analysis** - This analysis shows the flow of cash through the business. The new business begins operations with cash on hand from loans, grants, personal finances and other sources. On a monthly or quarterly basis revenues are offset by expenses leaving cash on hand to begin the next month or quarter.
- **Business plan documentation** - A documented business plan is essential whether or not outside funding will be sought. A typical outline will contain the following sections:
  - » Executive Summary
  - » Mission, Vision, Culture

>> Company Description
>> Opportunity Analysis and Research
>> Market Strategy and Plan
>> Management and Operations
>> Financial Analysis and Projections
>> Funding Request and Exit Strategy
>> Appendices

- **Business launch planning and execution** - Detailed plans to launch the new business need to be defined along a timeline consistent with available funding. The WBDC is prepared to offer whatever support the entrepreneur needs.
- **Business process design** - Business process design spans a business at any point in its life cycle. Before the business launches the operational business processes must be designed and their development and implementation scheduled into the launch plan. Many of these might be replicates of existing processes from similar businesses while others may require a creative twist in order to create barriers to entry for potential competitors or offensive strategies to protect the new business from the incursion of existing competitors. New business process design projects are important to the WBDC courses.

Business development projects

Once a business has been launched the business owner will encounter a number of development projects that could benefit from the WBDC program. These include:

- Business process design
- Business process performance metric design, implementation, and monitoring
- Business process improvement
- Problem solving
- Customized research projects

All five of these provide a rich source of projects for inclusion in the BIC project portfolio. The terms, conditions, and benefits of the Entrepreneurial membership can extend through the first year of operations for the new business. Again, the WBDC wants to minimize the barriers to using their services and increase the likelihood of a successful new business launch.

THE LEARNING ENVIRONMENT TO STUDENT ENVIRONMENT LINKAGE

The WBDC establishes a linkage to the student that is totally different than current educational delivery systems. It includes the traditional teacher-learner knowledge delivery but that is only a small part of the relationship provided by the WBDC Model. Figure 4.2 illustrates the WBDC relationship.

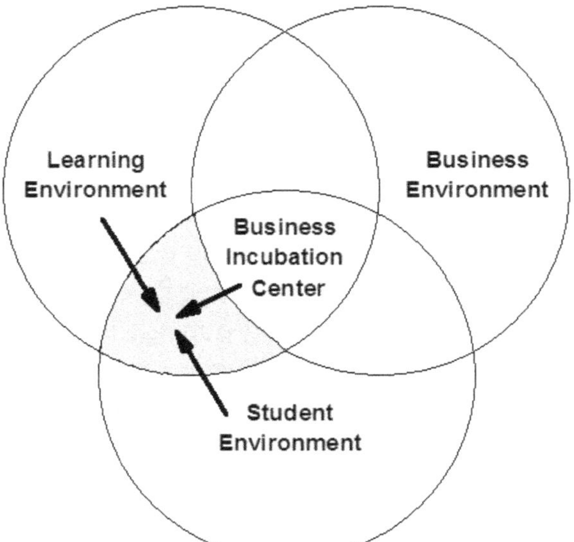

Figure 4.2: Learning Environment to Student Environment Linkage

The Student Learning Contract

The Student Learning Contract (SLC) is the agreement between the institution and the student as to the requirements of the certificate or degree program they are pursuing. The SLC is developed through a collaboration of the student and their faculty advisor. The Professional Development Plan (PDP) is an integral part of that collaboration. Chapter 6 discusses the PDP in detail. The SLC horizon may define a learning experience that extends over several semesters.

The SLC consists of three parts: Learning Objectives, Program Requirements and Courses.

- **SLC Learning Objectives** - Students will have some idea of where they are heading as far as their careers are concerned. With the help of their faculty advisor they can translate that into learning objectives. These objectives take the form of one or two brief sentences that identify the specific tools, templates, and business processes they will need to understand and know how to apply in order to meet those objectives and hence be prepared to occupy positions that define their career goal.

- **SLC Program Requirements** - With the assistance of their faculty advisor the student will translate Learning Objectives into a program and then into the course requirements for that program. There may be specific course required of a program as well as course groupings from which specific courses must be taken. And finally, some courses will specify learning objectives but leave options as to how those learning objectives are met.

- **SLC Courses** - Course requirements are not specific to the career goal but must be expressed in terms of specific courses and course content that will meet course requirements. To the extent that they can be known the specific courses are documented in the student's SLC. If there are unknowns, that is OK. They can be replaced with specific course content as they become known. Much of that content will be driven by the current inventory of BIC projects.

The initial SLC is signed by the student and approved by the faculty advisor. The SLC can be revised by mutual agreement of the student and their faculty advisor. The student drives the change process by updating their career goals. Updating the SLC can happen at any time but can only be effective at the beginning of the next semester.

Student team formation and objectives

Students will document their SLC through one-on-one sessions with their faculty advisor but they will execute their SLC in a team-based format. See Chapter 6 for an example of how this happens.

Team formation

Teams are formed around a common interest. That will usually be around a specific business. The BIC maintains a posting of the description of all active businesses and business ideas from which the student can choose and submit an application. They will have to appear before the current team and interview for their being invited to join. Since the business is an interest of the student they might be looking for a long-term relationship with that business and its team.

A student may also be interested in joining a team because of a specific learning opportunity it offers. By joining the team they will be able to fulfill a specific learning objective in their SLC. Once that objective has been attained they may elect to join another team for another learning opportunity.

Team objectives

The team will have aligned with a project from the BIC and that project will have certain objectives that its sponsor expects from any teams that might be assigned to their project. Students join this team because the business is of particular interest to them or because the project objectives align with their program objectives.

So the team is now tasked with meeting the project objectives as submitted by the business sponsor while paying attention to the individual student learning objectives. The tasks assigned to each team member must:

- Provide a learning opportunity for the student
- Contribute to one or more project objectives

This is a win-win situation for the student and the business sponsor. The student makes progress on her SLC and the business sponsor makes progress on his project. Furthermore the student is now motivated to move on to another learning objective and the sponsor can likewise move to the next phase of the project.

If the project is a business formation project, the motivation is heightened if the student is interested in the business as the direction they wish to take in their career and professional development. If the entire team has that same affinity to the project, this is the best of all possible learning and business formation projects. If the project is a business development project, the motivation is heightened if the student is interested in seeking employment in a similar business. Their experience with that business gives them a leg up on the competition for that position.

PUTTING IT ALL TOGETHER

So the foundation for all WBDC programs has now been laid. The relationship between the Learning Environment and the Business Environment is in place. Programs for participating in the classroom experience are documented and attached to the appropriate courses. Projects are documented and listed in the BIC for use in student learning experiences.

CHAPTER 5

The Business Environment

The Business Environment is the only meaningful resource for supplying the WBDC with learning program opportunities in the form of classroom participation and projects. It is a very rich source but has to be developed if it is to be used effectively. There are really two distinct Business Environments to consider as we discuss these opportunities: the Entrepreneur Environment and the Business Owner Environment. Each of these environments has unique and different contributions to make and projects to offer. In Chapter 4 the focus was on the pull coming from the learning community to gain the involvement of the business community in the learning program. In this Chapter the focus is on the pull from the business community to get the support and help of the WDBC in their business formation and business development efforts. The WBDC partnership with the business community can be a win-win partnership. In this chapter I share the details of that partnership.

THE BUSINESS ENVIRONMENT

Based on the trends we discussed in Chapter 2 technology is the center of all meaningful discussion of new business formation and business development. Technology can replace a number of business functions either totally or partially. In most cases that replacement means removing the human element and replacing it with systems that take responses directly from the customer without the need for an intervening person except in special needs situations. Because of this a number of jobs have been permanently lost. Call Center and Customer Service Center staffs have been significantly reduced or sent off-shore.

So while technology can be your worst nightmare it can also be your best defense. If the systems you develop are difficult or impossible to copy, you have created a barrier to entry in that line of business. Technology can also be used to solve problems that otherwise were insoluble and that also creates a barrier to entry for new competitors. The business owner also has to account for the trend in technology being integrated into business functions with many of the backroom functions being provided by outside vendors. Software as a Service (SaaS) is the coming reality.

The Entrepreneur Environment

The WBDC will partner with the entrepreneur from the first glimmer of an idea to the early formation of the business. This is a critical component of the WBDC Model because it gives students an opportunity to team with the entrepreneur and practice the formation of a real business in preparation for the day when they may choose to walk in the shoes of the entrepreneur.

The entrepreneur needs all of the help they can get and they typically don't have deep pockets. When I formed my business in 1990 I was in their shoes and know from personal experiences what they have to go through. In my case the early client engagements were used to bootstrap other products and services. The economy is very different now but many of the activities have not changed in the intervening 20 years.

To take advantage of the services discussed later (see Figure 5.2) the entrepreneur and the business owner should be WBDC member businesses. This will allow them to take advantage of most services at no or very low costs. The WBDC offers memberships to entrepreneurs, embryonic businesses, and established businesses. The membership fee is indexed on company size as measured by gross revenues. No revenues - no membership fee. That means that normal business formation, business process design and launch needs can be met within the WBDC with little or no cost to the entrepreneur. To the extent possible the WBDC minimizes the service and cost barriers to entry for new businesses.

To help the entrepreneur and the business owner, partnerships with student teams are formed through the WBDC and designed to benefit both parties by a type of value-added system. The student team delivers products and services (business value) to the meet the expressed needs of the entrepreneur and business owner in return for which the entrepreneur and business owner collaborate with student teams to help them meet their learning needs (learning value). Entrepreneur member benefits include a variety of consulting services offered by student teams. Routine services are offered pro bono while more complex services are offered for a significantly discounted per diem rate. The cost to the entrepreneur is generally not the exchange of money but instead the sharing of their ideas and experiences with student teams to support their learning needs. Business owners who are members of the WBDC can take advantage of WBDC services at reduced costs. Nonmembers can also take advantage of these services through a fixed bid contract with the BIC. The WDBC is a non-profit organization. Non-disclosure and non-compete agreements can be put in place at the request of the sponsoring manager.

For entrepreneurial members new business formation services

are pro bono. Since these pre-launch businesses have no revenues they do not have any membership fees to pay. The WBDC does everything it can to remove barriers to the entry of the entrepreneur into the BIC. Entrepreneurs and their small business is the lifeline to job creation and must be given every opportunity to grow. The entrepreneur will be able to utilize pro bono assistance from student teams. The return on the student's time investment is the opportunity for them to learn in a real world setting. For example, as part of their learning contract with the WBDC they can get involved in business formation research projects such as:

- Feasibility studies
- Competitor analyses
- Market analysis
- SWOT Analysis
- Strategic market planning
- New business process design
- Custom-designed research projects

The WBDC Curriculum offers a comprehensive portfolio of in depth business analysis and project management courses in response to the project requests of the entrepreneur. This provides the necessary knowledge and understanding of the infrastructure processes that support process design and improvement.

It is obvious that the learning opportunities are significant. The curriculum content reaches across the life cycle of a business process.

The above activities are the same as those that student teams will engage in as they test their own ideas for new businesses. The entrepreneur isn't getting the professional services that a consultant would offer but they aren't paying the consulting fees either. They need to remember that the students are in learning mode and are partnering with the entrepreneur for mutual benefit. It is expected that the exchange of ideas will have a lasting value on the student and the entrepreneur. They will all learn together.

The Business Owner Environment

The business owner is challenged just as the entrepreneur is challenged. The entrepreneur's challenge is to successfully launch a new business. The business owner's challenge is to survive and grow the business.

Starting a new business is risky and that is no secret. There isn't much data but I'll share what I could find. Figure 5.1 shows that a business that has survived until year "n" has a probability of about 0.82 of surviving until year "n+1". So that about half of the businesses that start in 2011, will still be around in 2014. That is an alarming statistic and every one of the entrepreneurs and business owners are convinced that they will be one of the businesses that survive for at least another year. How many of those that failed might have been saved through better planning and positioning? How many might have been saved by defensively and offensively leveraging technology for sustainability? How many might have been saved by protecting the business against the incursion of global competitors? How many businesses should have never launched in the first place? We'll never know the answers to these questions but what we do know is that continuous strategic planning is essential. In starting a new venture an entrepreneur will brainwash themselves into thinking that their business idea will be one of those that survive despite the risk. It's hard to face reality when you think you've got the best business idea since God invented dirt. The WBDC BIC is poised to cost effectively support the entrepreneur beginning before the business is even launched and continuing for as long as the entrepreneur gains value from their membership. The WBDC BIC and the programs it supports will mitigate member business risks as much as possible.

Percent
surviving

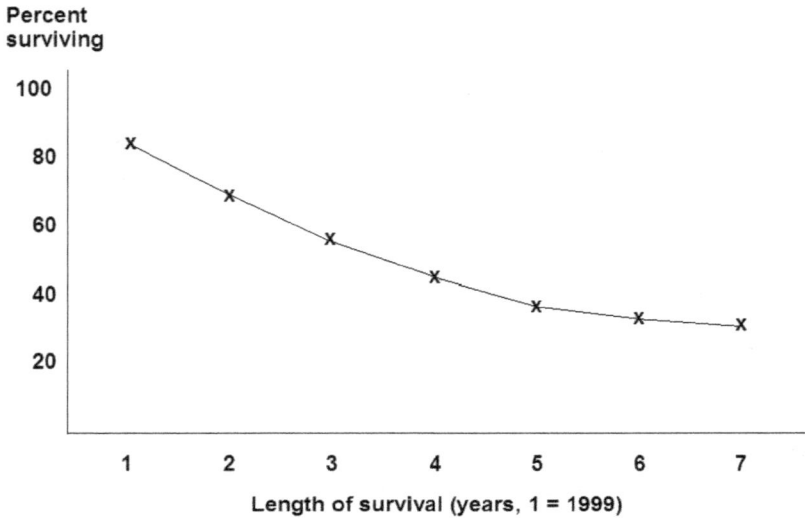

Figure 5.1: Business survival data

Adapted from data reported in: Knaup, Amy E. and Merissa C. Piazza, (2007).
"Business Employment Dynamics Data: Survival and longevity, II,) Monthly
Labor Review, September, pgs 3-10.

The business environment is the living laboratory and source of all practice in the WBDC Program. The WBDC offers a true integration of the Learning and Business Environments.

Local businesses are an integral part of the learning experience and contribute to the program in a variety of ways such as:

- They are a resource to the Learning Environment through guest lectures, panels, etc.
- They invite students to visit their company for tours and observation of operations and attendance at business meetings, problem solving sessions, process design, process improvement and a host of other situations.

- They may contribute equipment and other in kind services.
- They will mentor and advise the teams as they work in their business or in WBDC-owned businesses.

In the previous two chapters we discussed how the WBDC offers development opportunities for both workers and entre-preneurs. In this chapter let's turn our attention to the business owner and their business development challenges and see how the WBDC can benefit both the formation and improvement of their business processes.

The WBDC offers memberships to established businesses. The membership fee is indexed on company size as measured by gross revenues. Nonmembers can also avail themselves of these services at normal WBDC per diem billing rates. The WDBC is a non-profit organization housed in a post-secondary institution.

THE PROCESS OF BUSINESS FORMATION AND DEVELOPMENT

There is a business formation and development life cycle that describes the linkage between the business and the WBDC over the life cycle of the business. Figure 5.2 illustrates the business life cycle model that frames the WBDC support over that life cycle.

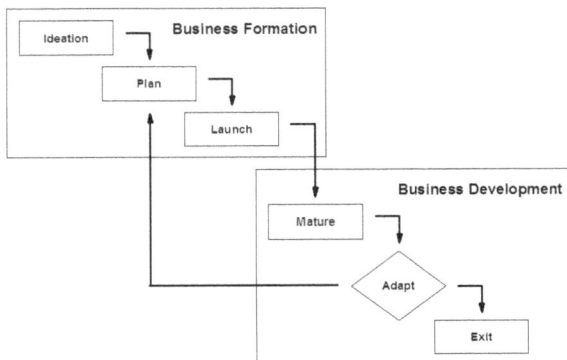

Figure 5.2: Business life cycle

Each phase in the business life cycle is supported by the WBDC as briefly described below.

Business Formation

Business formation starts with an idea. If the idea passes muster for feasibility and soundness it moves to a planning stage and finally to the actual launch of a new business.

Ideation

This is a comprehensive and objective look at the business idea presented by the entrepreneur. They will of course feel that if they build it and offer it, the customers will come. How could the customer not love what is being offered? So the first question is whether or not it is feasible. That translates into the skills, experiences, strengths, and weaknesses you have relative to what is needed, and how it will be financed. There are always questions about the market and who is in that market. The interest is on whether or not there is room for another competitor.

The WBDC offers the following services to aid the entrepreneur in the Ideation Phase:

- Feasibility studies

    The question to be answered in this section is: "Does it make sense to go forward with this business idea?" The question is answered by conducting the following:
    - » Define business process model
    - » Define and price services
    - » Revenue & expense budget
    - » Cash flow analysis
    - » Establish product/service matrix by competitor
    - » Write business validation report

    To the student team each of these defines a project that can be used to drive their learning experience. Some of them might fit exactly into an existing course. Others may require a learning plan to be developed and approved by their faculty advisor.

- SWOT Analysis

An analysis of strengths, weaknesses, opportunities and threats (SWOT) is major input to deciding whether or not the business idea has a chance. If technology can offer any leverages, this is the place to introduce that notion.

Figure 5.3 illustrates the situation quite clearly.

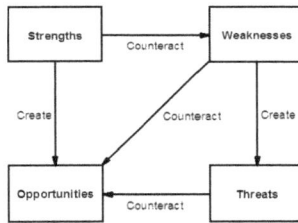

Figure 5.3: SWOT Analysis

Weaknesses and threats counteract opportunities. So the big question becomes: "How can the business idea be formulated so as to leverage technology to mitigate these counteracting forces?"

- Market Analyses

There are several questions that need to be answered with respect to the current market. For example:

» How big is the market?
» Will your product/service expand this market?
» Are market needs being met with the current offerings?

- Competitor Analyses

So you now understand the market so the next set of questions deal with how that market is served by the existing competition. For example,

» Who are the competitors and what is their market share?
» What are the strengths and weaknesses of the competition?
» How does your product/service offering fit into the current market?
» How do your strengths and weaknesses fit the market?
» Is there room for another competitor?

Plan

The plan addresses a variety of activities. Marketing and sales planning, organizational design and the business model, staffing, and, financial analyses are all required to launch the new business. The WBDC offers: Business Planning, business process design, cash flow analysis, and implementation planning.

- Business Planning

     There are several templates for the business plan. They all contain about the same information. A typical outline (Abrams, Rhonda, 2010, "The Successful Business Plan: Secrets & Strategies, 4th Edition," Palo Alto, CA: the Planning Shop, ISBN 978-0-9669635-6-4) might be:
     » Executive Summary
     » WBDC Description
     » Industry Analysis & Trends
     » Target Market
     » Competition
     » Strategic Position & Risk Assessment
     » WDBC Marketing Plan & Sales Strategy
     » Operations
     » Technology Plan
     » Management & Organization
     » Community Involvement & Social Responsibility
     » Development Plan, Milestones & Exit Plan

- » The Financials
- » The Plan's Appendix
- Business Process Design

   The new business may be able to adapt existing business processes from the industry. Even with that it is critical that the entrepreneur look for ways to leverage technology in the design of their processes.

- Cash Flow Analysis

   This is a simple analysis to construct. For input you will need the monthly estimated revenue and expense budget. It is expected that these numbers will increase over the planning horizon. The analysis begins with the initial funding as the cash position at the start of the business. In the first month there are estimates of revenues and expenses so that an estimated of cash on hand at the beginning of the next month is known. In the beginning the cash on hand will probably decrease for the first few months and will not begin to increase until sales begin to grow and revenues will exceed expenses. This calculation is repeated every month for at least the first five years of the business. As you move to the outer months of the analysis the estimates become less precise but are still useful. As the months pass the estimates are replaced by the actual and the analysis updated.

- Implementation Planning

   Planning for the launch of your new business is often constrained by the budget and implementation will of necessity be a phased implementation. New businesses are often bootstrapped by early business activity. Having one or two clients at launch is a common bootstrapping strategy.

Launch

An implementation plan will have been developed with the help of the WBDC.

BUSINESS DEVELOPMENT

Business development consists of a series of activities designed to mature the new business and sustain it through market dynamics.

Mature

The growth of the business compared to the performance metrics is monitored. Adjustments are made as required. The WBDC offers: defining performance tracking metrics, problem-solving, and process improvement.

- Performance tracking metrics

    Someone once said: "If you don't measure it, you can't manage it and if you don't manage it, it won't happen." The metrics must be defined up front before there is any business activity. The usual indicators of business are sales, cost of sales, number of leads, lead conversion, etc. But for the new business there are some other metrics that should be defined. All of the business processes are new and so are the people who use these processes. The big question is: "How are these processes performing?" The metrics should track time, cost, yield, error rates, etc.

- Problem-solving

    All of the performance metrics should have tripwire criteria such that if a value falls above or below the tripwire value, that triggers some form of corrective action to mitigate the situation and restore process performance to the nominal level. The corrective action puts a problem-solving process in place.

- Process improvement

    Closely related to problem-solving is a continuous process improvement process. Resolution of a performance issue may require an iterative approach.

Adapt

The world doesn't stand still because you are trying to grow your business. Markets change, technology advances at an unrelenting pace, and competitors come and go in the marketplace. In order to help businesses sustain themselves under these conditions the WBDC offers: strategies for leveraging technology for product improvement, exploiting technology to create barriers to entry of the competition, and using technology to increase revenues, avoid costs, and improve services. These are discussed below.

- Leveraging technology for product improvement

    A new technology is introduced into the business world. If there is an application to your business and you don't exploit it, your competition will and you will suffer the consequences. Integrating technology into a product that heretofore didn't utilize technology may create a new market. For example, greeting cards once consisted of a graphic and a poignant verse and nothing else. Along came the computer chip and all of a sudden the greeting card started acting like a computer, playing music or reciting a verse. If your business adopted the computer chip into your products, you probably found a new market and another profit center. Think of all the toys that exploited the computer chip and the new markets that that created. The technology has become even more pervasive with the introduction of voice response units and touch screens. There was a time when these technologies just came off the drawing boards and that was your opportunity to take advantage of them. If you waited, you lost. The WBDC will offer courses and other venues to explore creative and innovative ideas and applications for business advantage.

- Exploiting technology to create barriers to entry

    Just as technology can take jobs away technology can

also protect jobs from being taken away. To do so challenges the business owner's creativity. Finding ways to use technology in innovative ways to build a better service or improved product that cannot be easily replicated by the competition is the key ingredient to sustainability.

- Using technology to increase revenues, avoid costs, and improve services

    You have to be ever vigilant to your business and the processes that drive it. Becoming a lean, mean, fighting machine is a never-ending challenge. Technological change is constant and must be exploited wherever possible.

Exit

The WBDC has a limited offering here. The business owner will want to hire a professional to help them exit the market. Basically the business owner has to decide on how they are going to get out of the business and what the business is worth if they are selling it. This may apply to the entire business or only certain lines of business.

THE BUSINESS ENVIRONMENT TO LEARNING ENVIRONMENT LINKAGE

From the perspective of the Learning Environment this linkage is characterized by a pull from the learning community to gain participation by the business community. That linkage was discussed in Chapter 4. In this chapter the linkage works in the opposite direction. From the perspective of the business community this linkage is characterized by a pull from entrepreneurs and business owners to gain the support of the learning community. Figure 5.4, which is the same as Figure 4.1, can be used to describe this pull interpretation.

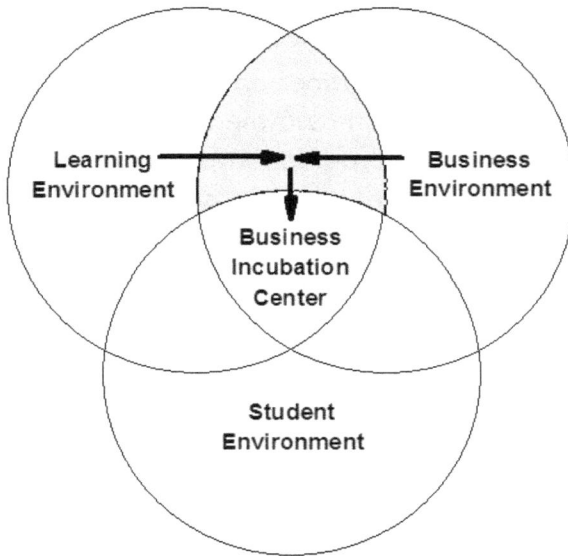

Figure 5.4: Linkage between the Business Environment and the Learning Environment

This linkage is another part of the disruptive innovation. The linkage is a true integration of the learning environment and the business environment through the BIC. The WBDC collaborates with entrepreneurs and business owners to document their needs and post them in the BIC as a project for one or more student teams. Focusing on their chosen project student teams participate in business activities (for example, attending meetings, training sessions, getting involved with problem solving, process design, process improvement or customized research projects). Businesses participate in classroom activities (attending team presentations, making presentations and delivering training). This proactive model supersedes what you may be used to for business involvement. Here business involvement is a critical part of the learning model while in typical applications business involvement may be little more than window dressing added to a course and not really an integral part of the learning experience.

The BIC plays a critical role in building the linkage between the entrepreneur/business owner and the Learning Environment. All entrepreneurs and business owners have constant challenges and problems to solve in order to form and grow their businesses. Many of these can be met using the services available from the WBDC through the BIC.

So, in effect, this linkage is defined by a reaching out from the entrepreneur and business owner to the WBDC for help solving a business formation or business development problem. These problems can be of any type from systemic (like defining a business) to isolated process improvement problems (like improving the yield of a process step). The entrepreneur or business owner documents their problems in collaboration with the WBDC and posts them in the BIC portfolio for consideration by one or more student teams. The contents of the BIC portfolio are continuously changing and provide a rich and dynamic source of projects to populate the curriculum.

THE BUSINESS ENVIRONMENT TO STUDENT ENVIRONMENT LINKAGE

This linkage delivers a quantum leap in benefits to the entrepreneur and business owner and to student teams as well. The WBDC BIC is a living laboratory for student teams to learn by doing projects for entrepreneurs and business owners. The BIC brings the project-driven curriculum to life. The Student Environment is team-centric and driven primarily by projects for business formation, process design, process improvement, and problem solving. Every one of these services can generate a variety of projects which are the fuel that powers the WBDC curriculum. So entrepreneurs and business owners are the lifeblood of the BIC project portfolio. New business ventures as well as existing businesses enrich the curriculum by proposing projects for inclusion in the BIC that the WBDC aligns with a specific course or courses in the curriculum.

To provide a public forum to encourage the growth and development of this relationship the WBDC has established an Ideation Café where students and entrepreneurs can come together to share ideas and lively conversation about these ideas. Think of it as a brainstorming session with food for the brain and food for the body. The objective is to test the feasibility and sustainability of the idea or some variant of it and to solicit support or partners. The entrepreneur holds the expectation that the WBDC might take on the idea as a business formation project and integrate it into the project portfolio of the BIC to further investigate the idea with the involvement of the sponsor and a student team.

The entrepreneur/student partnership is limited only by its own creativity

The BIC will always be the meeting ground for the student and the entrepreneur. The best situation arises when an entrepreneur submits a business idea that is of interest to a critical mass of students. Perhaps they have a similar business idea. From this common interest a unique partnership will develop as the student team idea and the entrepreneur's idea converge and improve as a result. This motivates the students to learn more about the feasibility of their ideas and it also motivates the entrepreneurs to know that they have the support of a team of students whose interests parallel their own. The synergistic effect is obvious. If the entrepreneur is successful in launching their business, they will have a ready supply of talent that has been tested and is available to join their embryonic company first without pay as students but then as salaried employees during the business development stage. The students might even negotiate for an ownership share of the business. What an exciting opportunity and truly unique among learning environments.

So the WBDC through the BIC partnerships that are created will have helped both the displaced worker and the entrepreneur meet their short-term goals - clearly a win-win opportunity.

New business formation is a daunting exercise given the global economy and the pervasive nature of technology. The WBDC has a program open to the public to ease some of the anxiety and hopefully give rise to new ideas that have a chance of success.

Entrepreneur involvement

This linkage integrates a learning experience and a business formation practice experience. The needs of the entrepreneur drive the need for learning on the part of the student team. In this linkage the student is participating with the entrepreneur in a project or problem submitted by the entrepreneur. Based on the student's skills and interests they may be involved in some aspect of the project or problem. For example, suppose the project involves process design. The student may be responsible for representing a suggested process design in UML format. Working sessions between student teams and entrepreneurs are charged with excitement. It is a place where student teams and entrepreneurs are brought together for mutual benefit. Both student teams and entrepreneurs are highly motivated by being able to collaborate and share ideas on a creative project.

THE BUSINESS ENVIRONMENT ➤

Obviously the WBDC BIC is not your father's incubation center. Your father's incubation center offers passive support for embryonic businesses during their formative period. Many such businesses will have secured at least a first round of venture financing. Space, access to shared office equipment, a shared conference room, telephone support, a mail box, internet access, a shared receptionist and other support services are available for a fee. While there is some limited support of this kind in the WBDC BIC one major focus of the WBDC is to support the entrepreneur with new business formation services which include feasibility analyses, business validation, problem-solving, business plan preparation and other customized research services.

The support to the entrepreneur is provided by teams of students under the supervisory control of a faculty member and the entrepreneur. The teams tackle projects and problems previously submitted and accepted into the BIC by the entrepreneur. These projects and problems are then available to student teams to be integrated into their coursework and programs of study. At the request of the entrepreneur these services can be offered under non-disclosure and non-compete agreements. I like to think of the WBDC and entrepreneur as forming a team whose goal is the successful entry of the entrepreneur into a new business.

The WBDC BIC is a living laboratory for student teams to learn by doing. Their learning environment is team-centric and driven by business formation, process design, and problem solving.

To be more specific the BIC business formation services that the entrepreneur might need were defined in the Business Formation part of the Business Development life Cycle illustrated in Figure 5.2 above.

The WBDC BIC is poised to cost effectively support the entrepreneur beginning before the business is even launched and continuing for as long as the entrepreneur gains value from their membership.

Business Owner involvement

These learning experiences involve business owners and student teams in business development projects. In the business formation role the services of the BIC are pro bono to the entrepreneur. They agree to work with student teams under a nondisclosure agreement. A business faculty member is assigned as learning advisor to the team and the entrepreneur functions as business advisor to the team. Students will naturally be attracted to the business idea and as gratis employees now they will often become salaried employees should the business be launched.

In the Business Development phase of the business life cycle student teams work with business owners on process design, process improvement, problem solving and customized research projects.

For the entrepreneur the linkage through the BIC creates a powerful partnership. The worker, as a member of a student team, partners with the entrepreneur for the purposes of business formation and business development.

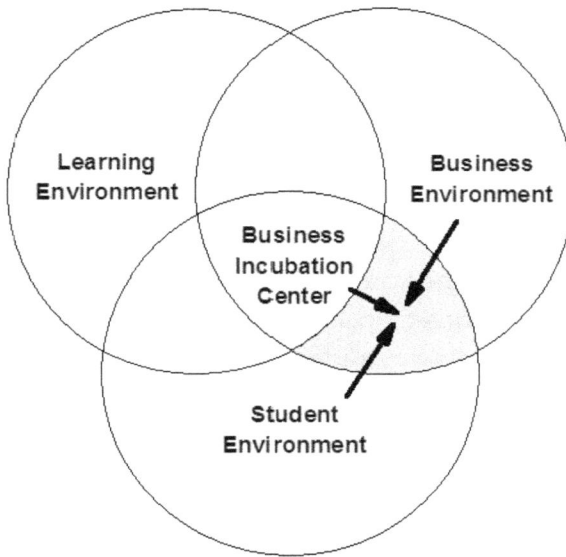

Figure 5.5: Business Environment to Student Environment Linkage

Similar to your father's incubation center the WBDC BIC offers passive support for its member businesses. Space, access to office equipment, a conference room, telephone support, a mail box, internet access, receptionist and other support services are available for a fee to member businesses. While there is some limited support of this kind in the WBDC BIC the major focus of the WBDC BIC is to support existing business operations with business process design, process improvement, problem-solving and related content areas. This kind of support is included in the membership fee and is provided by teams of students under the supervisory control of a faculty member and the guidance of the business owner. The teams tackle projects and problems submitted by local businesses which they integrate into their coursework and programs of study. At the request of the business owner these services can be offered under a non-disclosure non-compete agreement.

The WBDC BIC is a living laboratory for student teams to learn by doing. Their learning environment is team-centric and driven by business process design, process improvement and problem-solving. Every one of these services generates projects which are the fuel that powers the WBDC curriculum. The WBDC BIC is the meeting ground of local business owners striving to improve their market position and performance. Both student teams and local businesspersons enter into proprietary and confidential agreements to work together and benefit from the synergy created by their innovative partnership. In other words the WBDC BIC is an eclectic environment where ideas and problems abound and solutions are waiting to be discovered. Working sessions between student teams and business owners are charged with excitement. It is a place where student teams and business owners are brought together for mutual benefit.

The WBDC BIC is a living laboratory for student teams to learn by doing. Their learning environment is team-centric and driven by business process design, process improvement and problem-solving.

PUTTING IT ALL TOGETHER

You should now see how the WBDC Incubation Center is really the glue that links businesses to the classroom utilizing the BIC for the driver of the learning experience and the application of learning. All three linkages are facilitated by the BIC.

CHAPTER 6

The Student Environment

The WBDC is unique in that it provides an integrated environment for knowledge acquisition and practical experience. That is made possible by the creative design of the four components of the WBDC. Through the BIC the Learning Environment and the Business Environment are linked in a partnership that works for the mutual benefit of the student, the entrepreneur and the business owner. To my knowledge there is no program in the U. S. that can make that claim. Remember we are dealing with a disruptive innovation with the WBDC. The potential and impact of the WBDC is yet to be discovered.

The student should think of the WBDC as a one stop shopping experience for all of their career and professional development planning. Every student will have the opportunity to select a faculty advisor and business mentor. Those two relationships will serve them well through their educational program days and into their days as a worker or an entrepreneur. That relationship can be a lifelong relationship.

In this chapter we'll explore how the WBDC supports the student. Support can come on the form of preparing them for entry or re-entry into existing businesses or preparing them to be entrepreneurs and launch their own business. Both relationships with the WBDC can be lifetime relationships.

THE STUDENT ENVIRONMENT

The student enters the WBDC Program for two reasons:

- *Preparation for entry or re-entry into the job market -* Here they will work with a faculty advisor to describe their short and long-term career and professional development needs. Those needs are documented in a Professional Development Plan (PDP) that the student develops under the guidance and approval of a faculty advisor. That document is the input data used to create and update a Student Learning Contract (SLC). Several SLCs might be needed to satisfy the program requirements for a certificate or degree.

- *Formation support for a new business idea -* Student's who want to develop their own businesses will range from those who are new to the job market to those who are very experienced. They want to know if their business idea makes sense or how to modify it so that it does make sense. Assuming it does, they will then need all kinds of support for designing, developing, and planning for the launch of their business.

The typical WBDC student will be:

- from anyone of the six markets defined in Chapter 1
- sensitive to the possibility that jobs will be replaced by technology or moved off-shore
- looking for careers that are not easily outsourced
- seeking a program to prepare them for entry or re-entry in the job market
- evaluating the sustainability and feasibility of their new business idea
- seeking a program and support services that will prepare them and their new business idea for implementation

CAREER AND PROFESSIONAL DEVELOPMENT PLANNING

In order to qualify for senior level positions you will have to plan for them. Don't expect to get these promotions just because you are a nice person and happened to be hanging around the office when the opportunity arises. It can happen but it's like investing in the lottery. Since you own your career no one is going to do this for you. You will have to develop and maintain a plan that will result in your career and professional development targeted to your goal career position or the successful launch of their dream for a new business. That requires a concerted and diligent effort with the support of one or more mentors. I'll call this plan a Professional Development Plan (PDP).

A good PDP will answer the following questions:
> **Where am I?**
> **Where do I want to go?**
> **How will I get there?**
> **How well am I doing?**

The PDP planning horizon usually spans a full year with reviews on a quarterly basis or as needed. When used in a job situation the temptation is to schedule a PDP planning session in conjunction with an annual performance review process. That is not a good idea. In practice it is usually better to keep PDP development/ updating and the performance review processes six months out of phase with each other. Conducting a performance evaluation and career planning session concurrently is strongly discouraged. The performance review is not safe harbor for the individual. In a performance review they are under pressure and often put in a defensive position. In a career planning session they are in dream mode and don't need any pressure imposed on them from some other conflicting activity imposed upon these sessions.

The structure of the PDP process is portrayed in Figure 6.1. First note that the PDP process is never ending. Well I suppose it does end but that will be when you are dead or earned so much money that you don't care anymore. Note that the plan is defined by doing nothing more than continuously answering the same four questions. Between each round of answering the four questions your professional life can take several turns that justify reconsidering your answers from the previous round.

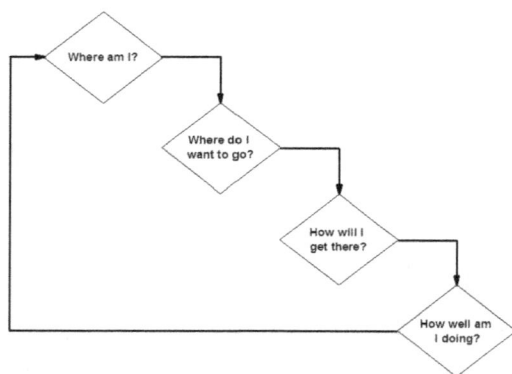

Figure 6.1: The PDP Planning Process

A PDP is the heart of every career and professional development effort. Your organization has a major role to play in PDP implementation and support and that is discussed in the next section. If your employer doesn't have such a process, don't use that as an excuse for not doing it. If I worked for such an employer, I would seek out my manager to work with me on the PDP and to be my coach and advisor. I keep telling my mentees that their employer owns their job but they own their career. Burn that idea into your brain and never forget it. If you don't keep this in mind but default to your employer to take care of your career growth, you will be making a big mistake. Employers have already shown their lack of interest in loyalty to their workers so don't expect that to change anytime soon.

So you should also seek out one or more professionals to be your mentors. Typically these will be people who have achieved professional goals not unlike your own. I remember that several years ago while I was still in my 30s that my primary mentor lived across the country and I only saw him once or twice a year even though we stayed in contact by phone and snail mail (the internet was still someone's dream). Mentors can change over time and you can have more than one at a time. Who are the people you admire and why do you admire them? Answer that question and you have a good start on recruiting your mentors.

A PDP should contain both a short-term and a long-term career goal. The short-term career goal should cover the next planning horizon - most likely one year. It will be a detailed description of what you have planned to do in order to achieve your short-term goals. The long-term career goal will be defined but at a less detailed level.

By way of introduction I think that a good PDP is a plan developed by you and your coach/mentor/manager and consists of the four phases as briefly defined below.

Where am I?

In other words:
- What position do you occupy?
- What is your recent assignment history?
- What is your skills profile?

This is your resume and it puts you at a starting gate from which you can begin to put a career and professional development plan in place.

Where do I want to go?

While it may seem trite I like to ask everyone that comes to me for career advice "What do you want to be when you grow up?" Some of the answers are amusing:

- I don't want to grow up.
- Employed.
- I want your job.

In all seriousness this is a basic question that must be answered before any career plan can be built. The answer isn't a lifetime decision either. Rather it is the place you would like to be professionally based on what you currently know and understand and what the business situation is likely to be as your career and job preferences mature. You can change your mind every Tuesday if you like. The important thing is that you have a career goal. You know where you are and where you want to go and you can put a plan in place to work towards that goal. The goal can be very specific: "I want to be the CIO at Wal-Mart before my 50th birthday." Or it can be very general: "I want to be recognized by my colleagues as a thought leader in my profession."

The process for answering the question is through a relaxed conversation with one of your mentors. Your manager should be one of your mentors. Your manager is your best source for understanding where the opportunities lie for your short-term career prospects. And you should seek them out for that advice. Longer term prospects are best commented on by those who have a strategic perspective on the future and where the staffing opportunities are likely to be found. One of your mentors might be a person who occupies the position you seek as a career goal. You should seek their advice for that.

What if your short-term career goal is to have your manager's job? A good manager would welcome that news. When it is time for their promotion they would like to have a subordinate ready and willing to step into their position so they are free to take that promotion.

What if your long-term goal is to work for another company in a senior executive position? Again a good manager should welcome that news too. They will have plenty of opportunities to find out what they need to do to remove the reasons you want to grow your career somewhere else. If there is a way they can correct the situation and you decide to stay, they know they will have a loyal executive manager.

Both your manager and any senior level manager whose advice you seek should be able to translate your career goals into a position class or sequence of position classes that form a career path eventually leading to your ultimate career goal.

Your employer owns your job.
You own your career.
Don't ever forget that.

How will I get there?

At this point you know where you are (the position class you are in and the specific position in that position class that you occupy) and where you want to go (the position class of your career goal). So your next step is to build a career path for getting there. In fact you could build several career paths that all lead to the same destination. One of those many career paths will be the one you pursue in greater detail. That is what the PDP is all about.

And that is the answer to this question. The answer however will be given one step at a time. That step will be the PDP for the next planning cycle - probably 12 months. In preparation for that 12 month period you will have put a four-part PDP together.

The PDP consists of:
    Experience Acquisition
    On-the-job Training
    Off-the-job Training
    Professional Activities

They are each described below. Typically you will have reviews of the progress you are making. Those are often quarterly review sessions with your manager. With that thought in mind the PDP should be very detailed for the next quarter and have lesser detail for the outer quarters.

Experience Acquisition

This part of the PDP describes the acquisition of further experience mastering the skills and competencies needed in the current position. The further experience is also related to qualifying for entry into a higher level position in the same or a comparable position class. There may be certain areas that should be the focus of that additional experience - perhaps an area of noted weakness that needs improvement. So this is a "do more of the same" part of the PDP with the above considerations taken into account.

Here are two examples of Experience Acquisition as they might be represented in the PDP

- Seek out project assignments that have more of a business analysis focus than you have been doing.
- Support professionals who are more senior to you and have a business analysis skill that you need to improve to better meet current position requirements.

On-the-job training

First of all training does not imply attending a formal course.

Training can be very informal. For example, just attending an activity facilitated by an expert in the skill area of interest and observing what happens. Watching how someone does something is a learning experience.

On-the-job training is undertaken to increase the proficiency of skills and competencies needed to improve performance in the current position. If performance in one or more skill areas is below a nominal performance level, then training may be required to bring that performance up to acceptable levels. Training need not be expensive and certainly does not have to be time consuming. There are several opportunities right under your nose. For example,

- Offering to help a colleague with one of their tasks to improve your skill to perform that task on your assignment
- Attending a workshop to improve a current skill you are using on your job
- Volunteering to join a project team that will further challenge you to take your skills to the next level

You just have to pay attention and be on the outlook for them. For example, John is a Business Analyst (BA) Manager and has just been assigned to manage an Order Entry/Fulfillment Process Improvement Project. John has had limited experience managing improvement projects and he wasn't too confident he could handle this project. There had been several projects in the past to improve the Order Entry/Fulfillment Process with little success to report. John's career goal was to become a Senior Project Manager (PM) and eventually a Consultant to the Order Entry/Fulfillment Process. So John asked Michelle, a Senior PM Consultant to help him improve his project management skills from the perspective of the BA on a process improvement project. Here are two examples of On-the-job Training as they might be represented in the PDP:

- Look for opportunities to observe and support the project management work of a BA who wishes to become a Senior PM

- Take training programs (on or off site) to enhance the project management skills required of your current position

Off-the-job training

The purpose of off-the-job training is to increase skill proficiencies to the level needed to qualify for the next or some future position in your career plan. In other words it is not relevant to meeting any current skill proficiencies associated with current job requirements.

For example, Harry is a BA Task Leader. His long-term career goal was to be a BA Consultant. His short-term career plan was to become a BA Associate Manager. Since his project management skills are very limited he needs to get started on them. He is good friends with Larry who is a BA/PM Senior Manager. Larry is respected among the BA community as one of the best project planners and Harry felt that Larry had a lot to offer him. Larry was more than willing to be Harry's mentor and also to help him learn project planning. Because he was so well respected for his project planning skills Larry was in high demand and always very busy. Harry approached Larry and asked him if he could help Larry out in some way in return for which he could observe Larry. To Larry that offer was a no-brainer and so he took Harry under his wing. Harry shadowed Larry and began learning project management from the ground up. When Larry retired Harry took his place as a BA/PM Senior Manager and one of the organization's best planners. Harry was now one step away from his long-term career goal. Here are two examples of Off-the-job Training as they might be represented in the PDP:

- Take courses (on or off site) to add business analysis skills that will be required by your targeted position in the PM/BA Associate Manager position
- Look for opportunities to observe and support a professional practicing the business analysis skills you will need in your targeted position.

Professional activities

This part of a PDP addresses participation in activities that are not necessarily related to any current or near-term career goal. Rather they are just experiences that relate to your long-term professional interests. They might be related to better understanding your company's business. For example,

- Reading the literature on your profession or your company's lines of business
- Better understanding your competitors and how you might leverage your project management and/or business analysis expertise to improve your company's market share
- Involvement in a related professional society at the national and chapter level
- Conference attendance and networking with other professionals

Here are two examples of professional activities as they might be represented in the PDP:

- Read books and journal articles on topics relevant to your targeted position
- Attend meetings and conferences offering seminars and workshops relevant to your targeted position

How well am I doing?

Your PDP should be written so that it can be used to track progress. Just like you would develop an acceptance test in the form of a checklist so also should your PDP performance tracking be in the form of a checklist. These checkpoints should be held quarterly with your career advisor and on an as needed basis as well.

In addition to tracking progress against the plan on a quarterly basis so also should you consider revising your plan on a quarterly basis. These could be as simple as taking advantage of an opportunity that wasn't known at the last checkpoint and which you would like to include in your PDP. Or they could be a little more comprehensive and focus on changing your career path or even your career goal. In Figure 6.1 that is the feedback loop to "Where am I?" and the process starts all over again.

The PDP should be updated at least annually. Even though it is an annual plan there will be unexpected events that suggest revisiting and perhaps revising the PDP before your annual update date comes about. For example:

- You get laid off
- Your job disappears
- Your company closes
- An unexpected opportunity arises

Given your new situation you may wish to change your career goal. The opportunity that gave rise to your current career goal might have disappeared or changed to the point that it was no longer attractive. Demand for the position might have changed or even disappeared altogether. For example, many programming jobs have been moved off-shore and not likely to return. So your goal of becoming a Director of Application Development is no longer as attractive as it was a few years ago.

Taking charge of your career

There's more than one way to skin the cat and if you aren't comfortable with using the PDP as described above, then here is another format that you might find useful. I've used it for several years now for my own career and professional development planning and it works great so I'm sharing it with you. I have used some form of PDP for my entire professional career. The four-part PDP was most useful early in my career while I found the format shown in Figure 6.2 most useful later in my career. Use it or lose it!

We've all heard of vision and mission statements. I've added tactical plan to that and so Figure 6.2 is another version of a PDP. The data given in Figure 6.2 is my actual 2010 PDP.

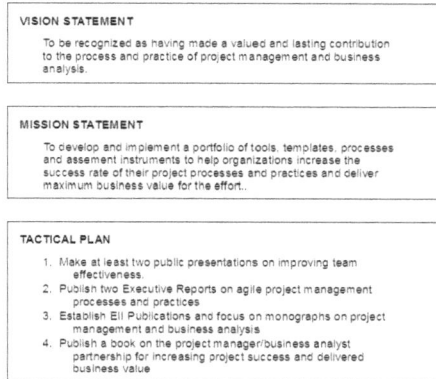

VISION STATEMENT

To be recognized as having made a valued and lasting contribution to the process and practice of project management and business analysis.

MISSION STATEMENT

To develop and implement a portfolio of tools, templates, processes and assement instruments to help organizations increase the success rate of their project processes and practices and deliver maximum business value for the effort..

TACTICAL PLAN

1. Make at least two public presentations on improving team effectiveness.
2. Publish two Executive Reports on agile project management processes and practices
3. Establish EII Publications and focus on monographs on project management and business analysis
4. Publish a book on the project manager/business analyst partnership for increasing project success and delivered business value

Figure 6.2: A Variation of the PDP

My Vision Statement hasn't changed in 20 years. I guess that means I figured out what I want to be when I grow up! The same can be said for my Mission Statement. I've changed the interpretation somewhat but not the statement. The Tactical Plan is a dynamic annual plan. I revise it in December of every year effective for the following year. So for 2010 I plan on making two public presentations. One presentation is done and the other is in planning. One Executive Report is finished and was published in May. The other is in preparation and will be published in the late summer. EII Publications has been formed and I'm just waiting on the legal processes to catch up with me before I go to the next step in my launch plan. The first product is planned for publication early in 2011. The book manuscript has been submitted to the publisher and will be published in October 2010. So I guess I'm on track to complete my Tactical Plan by the end of 2010!

The professional development services offered to the student through the WBDC are comprehensive and add value to career planning from the first entry level position through to senior management and business owner scenarios for as long as the student finds value in those services. In the following sections we will see how the linkages can be used for career and professional development.

Student Environment to Learning Environment Linkages

There are two linkages that affect the career and professional development of the student.

Linkage between the Student Environment and the Learning Environment

Figure 6.3 illustrates one of these linkages. During the time that the worker is a student in the WBDC Program their PDP will emphasize Off-the-Job Training because they are looking for growth into a new position. If they happen to be employed while they are in the program as a student they may have an On-the-Job Training Component and Experience Acquisition component but that would be less likely than simple Off-the-Job Training. So the courses that they take are focused on meeting certain knowledge requirements for their future position. Ideally they would like to have a practice component that is related to their career interests.

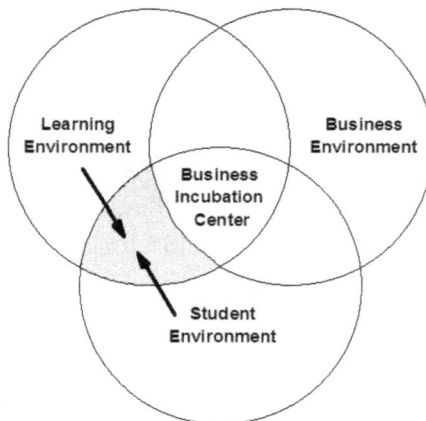

Figure 6.3: Direct linkage between the student and the Learning Environment

This linkage consists of the definition and delivery of programs, curriculum and courses. These are the building blocks of the SLC. They are adaptive in the sense that the student will have certain career and professional development interests and the programs, curriculum and courses are chosen and perhaps adjusted to satisfy those needs. As part of the application process the student can expect to meet with a faculty advisor to define their "Learning Contract." In effect this is an agreement with the WBDC and their home institution that when completed will result in the awarding of a certificate or degree by their home institution.

The BIC adds the application dimension to knowledge acquisition but it does so in a project/problem-driven curriculum. The projects/problems should be related to the courses they have in their program to meet Off-the-Job Training components.

When the BIC is used to facilitate the linkage between the student and the Learning Environment you have the situation depicted in Figure 6.4

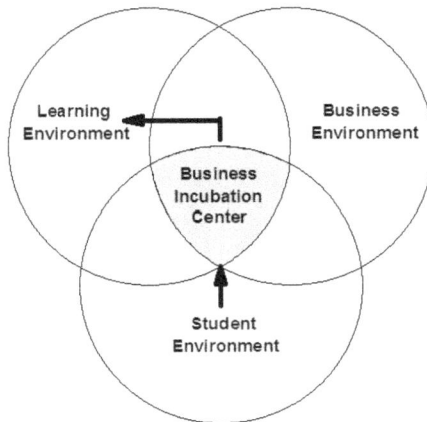

Figure 6.4: Linkage between the Student Environment and the Learning Environment utilizing the BIC

This is a dynamic linkage that takes projects from the BIC submitted by the member entrepreneurs and business owners and integrates them into student programs. That accomplishes two objectives. First, it establishes the project-driven approach characteristic of the WBDC Model. Second, it establishes a partnership of student teams to entrepreneurs and business owners.

STUDENT ENVIRONMENT TO BUSINESS ENVIRONMENT LINKAGE

These interactions are either part of a learning experience or an informational interview. As part of a learning experience the business will have a learning partnership with the WBDC. Students will be observing processes and perhaps gathering information relevant to a project or problem his team is working on for the business. As an informational interview the student may be seeking job opportunities as part of their PDP. Figure 6.5 depicts this linkage.

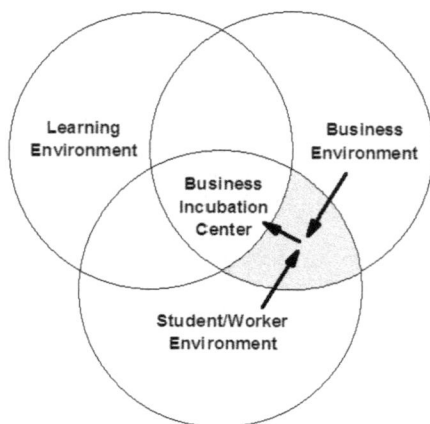

Figure 6.5: Linkage between the Student Environment and the Business Environment

The only linkage between the student and the entrepreneur or business owner that makes sense is one that utilizes the BIC. The BIC will offer several projects and problems that have been submitted by member businesses. These are all available to the student to support the project/problem driven aspects of their learning experiences.

This linkage is the heart and soul of the project-driven WBDC Model. The projects will range from business formation research to business process improvement and all range of projects, process design, process improvement and problem-solving efforts.

For every student team an intimate and confidential relationship with an entrepreneur or a business owner is an experience not otherwise equaled in any other program. This is where the rubber meets the road! Some students and workers will enter the WBDC Program with the thought of forming their own business rather than joining an existing business. Being able to work with an entrepreneur during business formation will be invaluable to the student.

Many will come with an idea but no detail as to how their idea will materialize or even can materialize but that is their motivation for entering the program. Their idea will become part of the BIC inventory and join many other ideas already in the inventory. This inventory becomes a rich source of projects for the learning environment. If a critical mass of students is interested in a business idea in the inventory of ideas, a team will form and a business formation project pursued as part of their project-driven curriculum. These ideas could also be suggested by a student as part of their application process or from a member business.

It's a win-win situation!

LAUNCHING A WBDC STUDENT TEAM

I thought it would be helpful to track a student team and their business partner through a typical launch of their project. These introductory activities are hypothetical as is the business.

The company is a 10 year old business called Lotta Barx Dog Food Company. Their owner/manager is Doc Sund. She is a successful veterinarian, a creative and aggressive businesswoman and is always on the lookout for other lines of business that complement existing Lotta Barx lines of business. Lotta Barx has no debt and has been profitable for the last 9 years.

In addition to the manufacturing and wholesale distribution to U.S. retail chains of their own selection of canned, dry and specialty dog foods, Lotta Barx offers boarding, grooming and trainings services to their local market. Those lines of business all operate at a profit. Lotta Barx has one discount store that is attached to the manufacturing plant. They opened it 2 years ago and it has just become profitable.

Doc has proposed to her partners that they launch a dog breeding program. Lotta Barx owns about 10 acres of raw land that adjoin the manufacturing plant and so space is not an issue. Doc hasn't researched the possible breeds and programs they might include. She does know that there are only 12 breeding businesses within a 50 mile radius of her company's location. She knows nothing about breeding but has customers all over the U.S. who do. With the tentative approval of her partners she has approached the WBDC BIC for help.

Coincidentally Scott Terrier, a recent college graduate and new student in the WBDC Program has always dreamed of starting a dog breeding program on completion of the entrepreneurial certificate program he has just started. Scott has shared his ideas with other WBDC students and has assembled a team consisting of Kay Neincorr, (Kay is soon to graduate from high school and works part-time in a pet store in a local shopping mall.) Ray Bees (Ray is a systems developer with over 10 years experience and recently lost his job when his company shipped all applications development to India. He loves animals and would like to find some way to make money working in animal care.) Del Mayshun

(Del was a customer services manager for a local auto dealership that went bankrupt and closed a few weeks ago. Del would like to find a job that utilizes his skills and can't easily be outsourced. Scott's dog breeding idea looked interesting to him so he signed on with the team for now.) So the team consists of 4 students with very different interests, skills and work experiences linked by a common interest. This will be typical of the WBDC student teams. The fact that they have an established member business that is linked to their business interests is an added bonus and will make their learning program even that more interesting and motivating. So with the advice and guidance of Doc the team decides in its kick-off meeting that the initial assignments to the team will be:

Scott: Conduct a major research project to identify potential breeds. Conduct a SWOT Analysis of the potential breeding program.

Kay: Investigate how to link existing lines of Lotta Barx businesses into the dog breeding business. Define and price service packages to offer. Do a competitor analysis including visits to all 12 breeders.

Ray: Begin defining and designing the information needs of the breeding business and business processes needed to support the new line of business and that can be integrated into existing business processes and systems. Consider using commercial application systems where appropriate.

Del: Identify the business processes that interact with clients of the breeding business. Begin designing such processes.

Team: Document a business plan.

The first thing to note is that every team member is assigned a task for which they are not appropriately skilled. That's OK because they have a task to complete and to complete it they will have to find courses and other learning experiences that will pre-pare them to complete their assignments. That is the nature of the project/problem-driven curriculum. First of all is their need to have a relevant PDP in place now that they have a business to work with. They would have a PDP already in place and may need to revise it. Let us assume that has been done.

PUTTING IT ALL TOGETHER

So the WBDC Model embraces the six student markets and their needs for career and professional development. More im-portantly, the Model reaches across a person's entire work life from first entry to retirement and provides continuing services as requested.

CHAPTER **7**

Design and Implementation Template

Implementing a WBDC Model in your market is a major project not to be taken lightly. This chapter describes a template that I recommend you use to create your implementation plan. For more detail on how to apply and customize this template see the accompanying publication: "The WBDC Model: A Comprehensive Guide to Design and Implementation" (Worcester, MA: EII Publications, 2011). An outline of the Guide is given later in this chapter. For information on how to order your copy, get a complete set of Word template files and establish an update subscription, contact me at rkw@eiicorp.com.

In actual application you will apply the template implementation plan to meet your specific customization of the WBDC Model. Every WBDC Model implementation is unique and so is the project to implement your WBDC. All I can hope to do in this book is share a template that will get you started.

A PERSPECTIVE ON IMPLEMENTING THE **WBDC** MODEL

While I encourage every institution to dream the dream my advice is to start with a realistic goal and let your implementation grow with time and market demands. That is the nature of a disruptive innovation and my recommendation to you. One approach is a phased approach. Beginning with the prioritized list of requirements, decompose them into two or more phases. Implement the first phase and track its performance in the market. The second phase will make adjustments to the first phase and add the next set of prioritized requirements.

I think it is a good idea to keep the high level look of the concept behind the WBDC Model constantly in mind. Figure 7.1 is a repeat of Figure 3.2. The most significant message in this graphic is the central role of the BIC in the model. It is, so to speak, where all the action takes place. Think of it as the clearinghouse for all WBDC activities and the linkage between any two environments. Without a fully functional BIC you can't implement the WBDC Model.

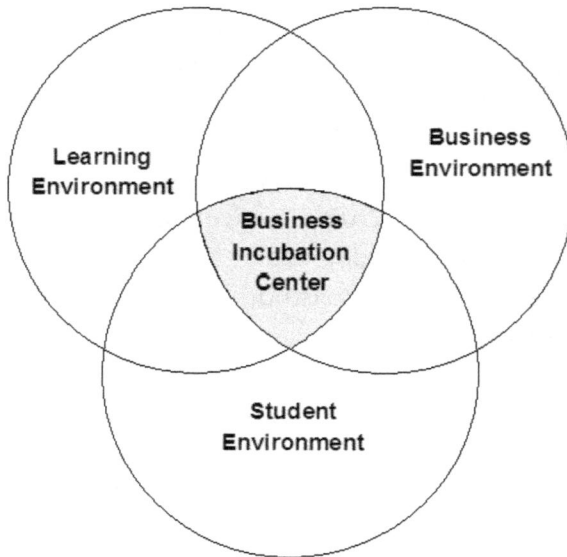

Figure 7.1: The concept of the WBDC Model

Recall from Chapter 3 that the WBDC BIC intersects and is part of all three environments.

- **The Learning Environment** - The WBDC BIC is a source of all project/problem-driven needs from the business members. It is the depository of all of the WBDC business ideas and fledgling businesses. It is the starting point for student teams and their learning experiences.
- **The Student Environment** - Student teams will work in WBDC businesses and use those businesses to generate project and problems that drive learning requirements.
- **The Business Environment** - The business environment uses the WBDC BIC as the clearinghouse for the process design and improvement projects and other problems needing solution. By posting these opportunities in the BIC they bring their needs to the attention of the student teams.

Without a fully functional BIC you can't implement the WBDC Model.

For some institutions implementing a fully functional BIC requires establishing a relationship with their local business community that heretofore did not exist. Some may have taken a small step in that direction by having internships and a guest lecture series but that is only a very small step. The WBDC requires establishing a real partnership with entrepreneurs and business owners. That will be a big step for some institutions. The Guide provides sound advice and a process for doing just that.

WBDC Model Design & Implementation Template

The transition from whatever model you are now using to a WBDC Model is significant. Figure 7.2 illustrates the template I recommend you use. If I were to consider implementing the

WDBC, I would spend considerable workshop time with the key players assuring they understand the scope of the undertaking and the fact that it is an adaptive effort. The initial version is just that - a version. I promise you that it will change but only experience with the initial version will give you the guidance for making those changes. Here is a place where an honest and open project scoping and requirements elicitation exercise is an essential ingredient for success.

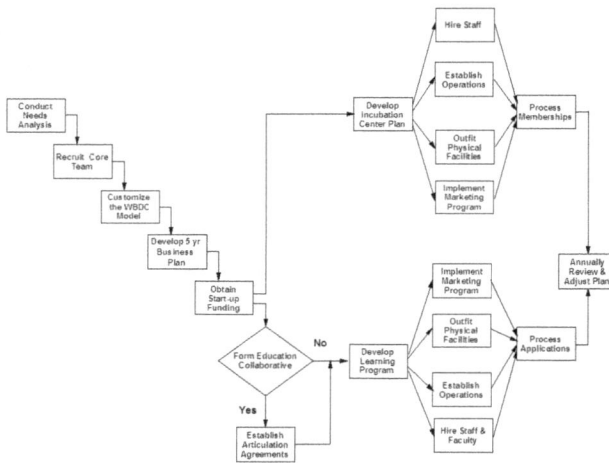

Figure 7.2: WBDC Design & Implementation Template

The Guide presents a more detailed implementation template forms and completed examples that can be further customized to your situation and needs.

Conduct Needs Analysis

The first step will be a business validation step in the form of a needs analysis. This needs analysis will answer the six questions discussed below.

If your design and implementation project doesn't embrace answering these six questions, it is incomplete. Go back to the drawing board. You will have to do better. Managing this project is nothing more than organized common sense. I am not advocating some complex approach that requires training and expertise to implement. If you just take the time to stop and think about what you are about to do, you will be on safe grounds. Keep the following six questions in mind and you will be well on your way to a successful design and implementation project.

What is your market situation?

The best way to answer this question is to use a Strengths, Weaknesses, Opportunities and Threats (SWOT) Analysis. The four parts of a SWOT Analysis are not independent of one another as illustrated in Figure 7.3. Figure 7.3 is a repeat of Figure 5.3.

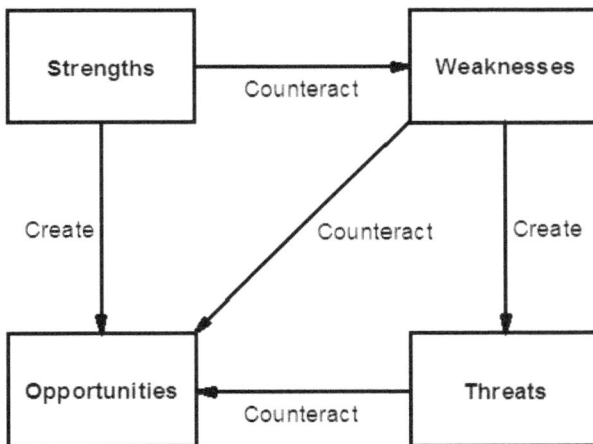

Figure 7.3: Dependency among the SWOT components

In completing a SWOT Analysis the team has to think in terms of uniqueness. For example, if you and your competitors all have a faculty with extensive business experience, you cannot list that as one of your strengths because all of your competitors also have a faculty with extensive business experience. At best you have a neutral position with respect to faculty business experience.

- Strengths

  There are two kinds of Strengths you need to consider. The first are temporal strengths. These are Strengths but only for the time being. All of your competitors can mimic temporal strengths and then they are no longer Strengths because all of your competitors have them. The second are permanent strengths. They cannot be copied by your competitors at least not in the short-term. Location might be considered one of your permanent strengths. Below is a list of things to think about as you build your list of Strengths.

- Temporal Strengths
  - » Professional degree breadth and depth
  - » Graduation production
  - » Price
  - » Capacity
  - » Access to local businesses
  - » Key faculty & staff
  - » Use of adjunct professional faculty
  - » Physical plant
  - » Library
  - » Market share
  - » Technology
  - » Positive reputation in the external community
  - » Free parking
- Permanent Strengths
  - » Location
  - » Alumni
  - » Program reputation
  - » Student retention
  - » Links between theory and practice
  - » Expertise in teaching non-traditional students

- » Faculty are open to changes in educational delivery models
- » Active business involvement in the learning environment
- » Positive brand awareness
- » Proactive faculty/business partnerships
- » Support for learning outside the classroom
- » Articulation agreements
- » Campus characteristics
- » Board of trustees commitment to professional programs
- » Openness to collaborative partnering for mutual benefit
- » Community respect for institutional leaders
- » Established SBDC
- Weaknesses

Weaknesses can also be temporal or permanent. A temporal weakness is one that can be removed with some effort. Permanent weaknesses are basically those that your organization inherited at birth and cannot be removed. Some examples of each follow.

- Temporal weaknesses
  - » Operating at capacity levels
  - » Physical facilities
  - » Faculty are not supportive
  - » Too much internal contention and politics
  - » Lack of cooperation across programs
  - » Planning is not responsive to community needs
  - » Fiscal uncertainty
  - » Faculty are change intolerant
  - » Website
  - » Unavailability of courses
  - » Community perception that programs are not aligned with needs

- » Underutilized technology
- » Outdated technology
- Permanent weaknesses
    - » Location
    - » Faculty retention
    - » Mission not aligned with professional development
    - » Excessive bureaucracy
- Opportunities

This one is difficult because the WBDC Model is new to every institution and most Opportunities would be held by all organizations and hence not an Opportunity to anyone. As Figure 7.3 indicates Strengths can create Opportunities so keep this in mind as you build your list of Opportunities. In any case, some things to think about include:

- » Expand existing project and problem based curriculum
- » Potential for workforce development programs
- » Increased demand for career and professional development and lifelong learning
- » Expand community outreach programs and communications
- Threats

As Figure 7.3 indicates Weaknesses can create Threats so keep that in mind as you build your list of Threats. Threats can be internal or external. Brief listings of each type are given below.

- Internal threats
    - » Loss of key staff to competitor institutions
    - » Ability to maintain existing capabilities
    - » Politically charged environment
    - » Priorities of the leadership
    - » Lack of focus
    - » Resistance to change

- » Traditions
- » Lack of customer service
- » Contention between training and education
- » Excessive competition for internal resources
- External threats
  - » Political environment
  - » Competing services offered by competitors
  - » Competitor intentions
  - » Budget crisis
  - » Decline in middle class jobs
  - » Emergence of new competitor institutions
  - » For profit and online responsiveness
  - » Federal government priorities

What do you need to do?

So the SWOT Analysis helps you understand what needs to be done to have an effective WBDC Model for your market. If you are lucky, you will be able to do everything on your list but to be more realistic maybe the strategy will be to do some of what you need to do. If that is the case, a good approach is to plan a phased implementation. That phased implementation might begin by prioritizing what you need to do. One prioritization rule might be to list what you can do according to the following, which is known by the acronym MoSCoW:

- **M** means must do
- **S** means should do
- **C** means could do
- **W** means wouldn't it be nice if we could

The early phases would implement the **M** and **S** items on your prioritized list. Then take a breath and assess the effectiveness of what has been implemented. Then you are in a position to decide what to do with the **C** list and finally the **W** list. For each item on the **C** and **W** lists your criteria for inclusion would be the incremental business value and social value having those capabilities would add to your WBDC Program and the community you serve.

What can you do?

Once you understand your market position relative to your competitors you are in a better position to decide what you can do. Of course you would like to meet all identified market needs but that may not be possible. A partial solution may be all that you can do now and leave it to the second version team to provide additional parts of the solution. There can be a number of reasons for only being able to provide a partial solution that are outside yours, the clients or the enterprises capabilities and resources.

How will you do it?

This is the beginning of your 5 year business plan. Describe how you intend to deliver an acceptable solution and work out the details in the business plan. The ideal business plan will be the cradle to grave description of the work to be done, how long it will take, what resources are needed and how much the solution will cost. Whatever you may have been taught elsewhere the reality is that developing such a complete plan is seldom possible. There are many situations that prevent the complete plan from being written and attained. When you can't build that ideal plan you will have to use some variant of a just-in-time planning model. A just-in-time plan actually evolves over the course of doing the project. Plan a little do a little and continue in that repetitive fashion until the project is completed. Here is where organized common sense makes its initial stand.

How will you know you did it?

Business reasons (success criteria) were put forth for justifying doing the project. An acceptable solution will meet both require-ments and the business success criteria. Client requirements are what the client believes defines the best way to meet those business success criteria. If satisfying client requirements does result in the best solution, then those business success criteria will have been met as well. These will be quantified metrics with specific values that define project success. Either you met them and the project was a success or you didn't and the project was to some degree a failure. The success criteria should be stated in such a way that it is obvious that they have or have not been met. This is not a debate that takes place at the end of the project. It is a quantified statement made with the client during the very early stages of the project.

How well did you do?

The project work is complete and the solution has been implemented. It's time for the post-mortem. There are two different things to consider in analyzing how well you did. The first is the quality of the product that was produced by the project. Did it meet the client requirements and did it achieve the business success criteria that justified doing the project in the first place? The client (and you as well) assume that by satisfying requirements the desired business value was achieved. The cause and effect relationship that you identified may or may not be the dominating factor. Perhaps there are other confounding factors that were not considered in the original requirements list. Requirements definition is complex. Most project management gurus would agree that in all but the trivial projects complete requirements definition cannot be done at the beginning of the project. Instead the requirements list is a changing and expanding list that evolves over the life of the project. Requirements are learned and discovered during the course of executing the project plan. Planning and managing such a project becomes a real challenge but is not insurmountable.

The second thing to consider is the process that was followed to produce the product. The correct analysis of the process that was followed will answer four questions:

- How well defined and documented were the project management processes you chose to use?
- How well did the chosen processes fit the needs of the project?
- How well did the team follow the chosen processes?
- How well did the chosen processes produce the results expected?

Answers to the first two questions will provide input to needed project management process improvements and an answer to the last two questions will provide input to needed practice improvement efforts (i.e., training needs or improved processes for making project assignments).

The contents of a Project Overview Statement

The last activity in the Needs Analysis is the WBDC Project Overview Statement (POS). The POS is your 2 minute elevator speech recorded on a one page document that describes the WBDC Implementation Project. I have used the POS since 1964. It has been an indispensible tool for project initiation in my consulting practice. It is a document that describes the project at a very high level and is written in a language that anyone in the organization who has reason to read it will understand it. So no techie talk is allowed.

The POS consists of the five parts described below.

- Problem/Opportunity Statement

What is the business reason for proposing this project? It might be a problem that so far has not been solved but is critical to the continued operation of the enterprise. It might be an untapped business opportunity. The most important aspect of the statement is that it is recognized by the enterprise as a problem or opportunity that does not need to be defended.

- Goal

What are you proposing to do about it? This should be a

clear statement of what you expect to achieve. It could be a partial or complete solution of the problem. It could be the first phase implementation (**M** and **S** lists) of the WBDC or a complete implementation. The decision is based on your confidence that you have identified all of the requirements of your WBDC solution.

- Objectives

As a further explanation of the goal what are your objectives in taking on this project? This will be a list of 6-8 objectives that form a necessary and sufficient set of objectives. That is, all of the objectives must be completed in order for the solution to be acceptable and none of the objectives are superfluous.

- Success Criteria

These are quantitative statements that clearly define what constitutes success. At the appropriate time they will either have happened or not. There won't be any arguments about which occurred. Success criteria are usually defined around increased revenue, cost avoidance or service improvement. Attainment of all of these success criteria is what defines an acceptable solution. That is, the solution delivers the expected business value that justified doing the project.

- Assumptions, Risks, Obstacles

What are the high level aspects of the project that might be obstacles or barriers to a successful outcome? These are noted now because they may be things executive management can mitigate.

An Example of the Project Overview Statement

The POS for a WBDC design and implementation project is given in Table 7.1.

Table 7.1: An example POS for a WBDC Implementation Project

POS Component	POS Content
Problem/ Opportunity Statement	Because of the weak economy workers in our market are having a difficult time defining a sustainable career and professional development program which includes preparing themselves to find employment in an existing business or launching their own business.
Goal	To design and implement a comprehensive service called Workforce & Business Development Center (WBDC) to serve the needs of workers and entrepreneurs in our market area.
Objectives	1. Design and implement a team-driven project-centric curriculum to serve the career and professional development needs of our post-secondary market over the life time of their professional career. 2. To be the resource for new business formation and business development in our community. 3. To establish a BIC that links the learning environment, the business environment and the worker environment for the mutual benefit of all three.

Success Criteria	1. The WBDC will raise $2M from grants and contracts by the end of the first year of the implementation project. 2. The WBDC will be self-supporting by the beginning of its 4th year of operation.
Assumptions, Risks, Obstacles	1. Raising $2M in seed funding is feasible. 2. No additional funding will be needed to achieve self-supporting status. 3. Faculty will be resistant to adopting a project/problem - centric curriculum model. 4. There are too many programs competing for resources.

The POS should be signed by the chief academic officer and the person who will manage the project. Until such time as the project plan is documented, the POS could change. Those changes must also be approved by the chief academic officer.

Recruit core team

The core team consists of those who will be on the project team from the beginning of design and implementation to the end. The core team is formed even before the project has been approved. Many of them should have participated in the Needs Analysis.

Core team members should be representative of and empowered to make commitments on behalf of the office or department they represent. The following offices and departments are typical of core team membership:

- Office of the Chief Academic Officer (Academic Vice President, Provost, or Dean of Faculty)

- Business Faculty (Dean of the School of Business, Chair of the Department of Business, Chair of the Department of Information Systems)
- Continuing Education (Dean/Director of Continuing Education, Dean/Director of Workforce Development)
- Chairperson of the Curriculum Committee
- Director of Faculty Development
- Director of Institutional Development/Advancement
- Director of Research & Development Office
- Office of the Chief Information Officer
- Director of Small Business Development Center
- Faculty who have expressed an interest in the WBDC curriculum design

A core team of 10-12 would be appropriate. If the core team gets much larger than 12 members, it may become dysfunctional. If size becomes a problem, the core team may evolve into executive level members and operational level members in the full project team. The operational level members will actually do the design and implementation work subject to the approval of the executive level members.

Customize the WBDC Model

Since you are about to begin a journey whose destination can only be idealized this is the time to define what you believe to be the WBDC Model you will implement. The template WBDC Model is your starting point and should be customized to your specific environment. The MoSCoW prioritized requirements list will be your best guide to the design of your WBDC Model. It will change even as you execute your business plan and project plan. So take your best guess and move forward. The recommended project approach should accommodate change so don't get too concerned about missing the target solution. You will learn about that solution as you do the project and can draw on early experiences using it. So the customization exercise is a good time to dream the dream!

Develop 5 year Business Plan

Using the customized WBDC Model a 5 year Business Plan should be developed. There are a number of good references for building this plan. A typical outline (Abrams, Rhonda, 2010, "The Successful Business Plan: Secrets & Strategies, 4th Edition," Palo Alto, CA: the Planning Shop, ISBN 978-0-9669635-6-4) might be:

- Executive Summary
- WBDC Description
- Industry Analysis & Trends
- Target Market
- Competition
- Strategic Position & Risk Assessment
- WDBC Marketing Plan & Sales Strategy
- Operations
- Technology Plan
- Management & Organization
- Community Involvement & Social Responsibility
- Development Plan, Milestones & Exit Plan
- The Financials
- The Plan's Appendix

The Guide presents a detailed template with selected example contents of the 5 year plan.

Obtain start-up funding

In most cases funding a WBDC implementation will require a multi-pronged and perhaps multi-phased strategy. A combination of support from local, state, federal, private and corporate programs will be required. Do not underestimate the scope and importance of this activity. It is in fact a challenging project in its own right. The 5 year Business Plan will include a financial plan that involves start-up funding. That could be provided partially by the host institution but most likely will involve outside funding or matching funds. Appendix A lists several public, private and corporate programs that have supported comparable programs. The Guide provides a more comprehensive annotated description of those and several other sources. Application information and strategies when appropriate and available are also included.

Form educational collaborative

There are two ways to proceed with the customization of the WBDC Model to a specific market. One is to go it alone in which case there is no need for an educational collaborative. This is not too likely or too smart a choice because it cuts the institution off from other institutions in their market.

The other is to form an educational collaborative, which is the more likely choice. There is a win-win strategy for those institutions that collaborate with other institutions in their market area. The collaboration involves deciding who will offer what programs and courses. A collaborative effort will raise all of the boats in the harbor. I can't state it any simpler.

There is a lot of advantage to having articulation agreements with neighboring institutions so that students from each institution can take advantage of programs and courses offered by the other institution that are not offered or not available when they need them. For example, there is a small private liberal arts college in my area that offers a minor in entrepreneurship. It has a strategic focus and is popular with students in a variety of BA programs. This college is the only one in the area that offers such a program. They often have problems getting a critical mass of students to register for the courses in the minor program. What if the college had an opportunity to enroll students from other colleges and universities in the area into their minor? The articulation agreements offered with the WBDC would provide just such an opportunity.

The Guide describes a symposium that the host institution might offer to their community to encourage competitor institutions to join the cooperative and work collaboratively for the common good.

The agreement between the WBDC and a post-secondary institution is called an Articulation Agreement. These are commonly done in most institutions especially when they are agreements between community colleges and 4 year universities. For the WBDC the articulation agreements will certainly identify transfer credit agreements but they have to go further. The articulation agreements reach into market sharing opportunities. One institution might be strong in one business application area and be the place where students go to get that education. But keep in mind that the curriculum model used by that institution must be compatible with the WBDC host institution, i.e., team-driven and project-centric. In other words, the articulation agreement is more of a collaborative agreement as to how each institution will contribute to the WBDC Model of the host institution.

Develop BIC Plan

This is a major component of the WBDC Model and will probably be a phased development plan. There are four activities to be done to complete the development of the BIC. My advice is that they be done in the order listed.

Hire staff

The 5 year WBDC Business Plan will include the BIC Plan. The staffing plan will include a director and several support staff. The first position to fill will be the Director of the BIC. Additional staff will be support staff. These positions are demand driven and should be defined by the Director and filled as needed.

Establish operations

The services that will be offered by the BIC should be defined in the 5 year WBDC Business Plan. This activity will describe how the services will be provided. The following are typical:

- Contact information
- Hours of operation
- Facilities, services and fee structure
- Room Reservation Process
- Need Request Process
- Need Request Application
- Service Level Agreement (SLA)
- Business membership terms and conditions
- Business membership application

Outfit physical facilities

The BIC physical facilities must support the following activities:

- General meeting room to host business presentations, conferences and other large group events
- WBDC-owned company meeting space
- WBDC Ideation Cafe
- Member company skunk works meeting space
- Student teams meeting space
- Internet, networking and other computer facilities to support the above activities

There will be one integrated facility to serve the needs of the entire WBDC Program. The above list identifies the needs of the BIC only.

Implement marketing program

The BIC Marketing Plan will have been defined in the 5 year WBDC Business Plan. This activity is connected to the kick-off meeting that announces the BIC to the community. It will be a multi-pronged program that typically includes the following activities:

- Email announcements
- Web-site availability

DESIGN AND IMPLEMENTATION TEMPLATE ➤

- Direct mail campaign to local businesses
- Public speaking events
- Entrepreneur and business membership campaign

Develop Learning Program

This is the second major component of the WBDC Model and will probably be a phased development plan. To begin this activity the programs, curriculum and courses need to be developed. Some of this work will be done concurrently with the four activities listed below. My advice is that they be done in the order listed.

Hire staff & faculty

Faculty is needed to develop the details of the programs and curriculum and then to do all course development work. The faculty administration can be put in place and the high level description of programs and curriculum can begin.

Establish operations

This is an academic program first, last and always so it must integrate with existing academic processes and practices of the sponsoring institution. The WBDC program design process must adapt to the changing and varied needs of each student. In fact, a student's program objectives must drive the actual curriculum and courses they will need. Many of their courses will not be in the academic catalog but are designed around the student and student team learning objectives. That means their faculty will play an approval role in program requirements. A minimum program requirement should be defined by the institution and a WBDC program built to meet that requirement.

Outfit physical facilities

The physical facilities of the BIC and the learning facilities are shared space. The focus of the WBDC facilities is the "Great Room". This is one large room capable of handling the following events:

- Conferences
- Workshops
- 4 Classroom(s)
- Business presentations
- Catered buffet and seated meals
- Information sessions
- Bulletin boards

Adjoining the Great Room are several breakout rooms that serve not only the conference and workshop needs but also provide student team rooms, WBDC -owned business facilities and member business "Skunk Works."

Implement marketing program

Marketing the WBDC Program is a multi-media program. At the heart of the program will be information sessions held at the WBDC facility. These are open enrolment sessions.

Process business memberships

There will be three types of business memberships: institutional, entrepreneurial and business.

Institutional memberships

These are high school and post-secondary private and public educational institutions that have articulation agreements with the WBDC. The terms and conditions of these memberships vary with each agreement. There is no standard agreement. Some of the variables that determine the fee structure, terms and conditions are:

- Reciprocal course enrolment registrations
- Collaborative learning programs
- WBDC joint management and administration

Entrepreneurial memberships

Entrepreneurial membership fees are annual fees based on gross revenues from all sources (i.e., venture capitalist funding, grants, contracts, etc.). For the most part this membership is open only to entrepreneurs who business idea is in the formation stages. They will engage the WBDC services through the BIC.

Business memberships

Business memberships are annual fee-based memberships based on annual gross revenues from product and service sales. These memberships offer existing businesses a number of services including:

- Process design
- Process improvement
- Problem-solving
- SWOT Analysis

There is no additional fee beyond the annual membership fee for using these services. The terms of the membership discuss the assignment of student teams to the provision of the requested services. The service is initiated by having the business owner submit an application describing problem and the needed services. A meeting is scheduled between the business owner and the student team that has expressed an interest in providing the service within the context of their learning contract. The business owner discusses the project in more detail and answers any questions. If the student team is interested and the WBDC approves, the project is defined and planning is scheduled. All of these activities can be done under the protection of a nondisclosure agreement if requested.

Process student applications

One application process and application form is used for all six target markets. As part of the application process the student is invited to submit any idea they have for a business they would be interested in starting or teaming with someone with similar interests. Admission decisions are made continuously and newly admitted students can start anytime.

Annually review & adjust plan

Any disruptive innovation has a life cycle. Its original formulation may be very different than its final formulation. All we can say at the front end where we are now is that we believe that the right direction is the direction we chose. We try it and some of it works some of it doesn't and so we strengthen what works and change what doesn't and the cycle repeats itself. The final solution cannot be seen at the start. We only have our collective wisdom to guide us at the beginning and then we have our experience to further guide our efforts. Hopefully we will converge on a solution that meets everyone's expectations.

A COMPREHENSIVE GUIDE TO THE DESIGN AND IMPLEMENTATION OF YOUR VERSION OF THE WBDC MODEL

"A Comprehensive Guide to the Design and Implementation of a WBDC" is a separate publication that will soon be available directly from EII Publications. It is a three-ring binder that can be continuously updated as new implementation information becomes available. For those who want to strike out on their own and design their own WBDC and implement it in their market, the Guide can serve as your template. Here is the major topic outline of the Guide.

    I.   Design & Implementation Overview

        A. An Overview of the WBDC Model

        B. Conduct a needs analysis

            1. Current programs

            2. Market

            3. Competition

            4. SWOT Analysis

            5. Where are you?

            6. What do you want to be?

            7. How will you get there?

II. Recruit Core Team
   A. Project Manager
   B. Administrators
   C. Faculty
   D. Institutional Representatives
III. Define your ideal WBDC Program
IV. Develop 5 Year Plan
   A. Executive Summary
   B. Mission & Vision
   C. WBDC Description
   D. Leverage strengths (from SWOT Analysis)
   E. Mitigate weaknesses and threats (from SWOT Analysis)
   F. Prioritize opportunities (from SWOT Analysis)
   G. Marketing Strategy & Plan
   H. Management & Operations
   I. Financial Analysis
   J. Funding Requirements
V. Obtain Seed Funding
   A. A detailed description of public and private sources
   B. Corporate donations
   C. Template proposal inserts
VI. Develop Incubation Center Plan
   A. Incubation Center Services
      1. WBDC-owned businesses
      2. New business ideas
      3. Process design
      4. Process improvement
      5. Problem solving
      6. Skunk works

B. Implement Marketing Plan
C. Outfit Physical Facilities
D. Establish Operations
E. Hire Staff
F. Process Memberships
VII. Form Education Collaborative
A. Advantages & Disadvantages of a Collaborative
B. Develop strategy
C. Identify collaborative members
D. Hold symposium
E. Draft agreement
F. Receive membership applications
VIII. Establish Articulation Agreements
A. High Schools
B. Community Colleges
C. Colleges & Universities
D. Corporate
E. Other
IX. Develop Learning Program
A. Program Components
1. Career & Professional Development Services
2. Project-centric and team-driven design
3. Core curriculum
4. Core course descriptions
B. Facilities
1. Great Room
2. Breakout Rooms
3. Media
4. Hardware
5. Software

    C. Memberships
- 1. Corporate
- 2. Institutional
- 3. Individual

X. Develop Marketing Program
    A. Public speaking
    B. Conferences &Exhibits

XI. Outfit Physical Facilities
    A. Great Room
    B. Breakout Rooms
    C. Product Demo Rooms
    D. Administration & Staff Offices

XII. Establish WBDC Operations
    A. Financial Aid
    B. Registrar
    C. Admissions
    D. Information Technology
    E. Career & Professional Development
    F. Business Development

XIII. Hire Staff & Faculty
    A. Executive Director
    B. Dean of the Faculty (member of Executive Team)
- 1. Associate Dean of Curriculum
- 2. Manager of Articulation Agreements
- 3. Faculty

    C. Director of Information Technology
- 1. Manager of computer labs
- 2. Manager of Product Demo Rooms
- 3. Systems Developer

    D. Director of Career & Professional Development (mem-

ber of Executive Team)
1. Counselors
2. Business Mentors
3. Faculty Mentors
E. Director of Business Development (member of Executive Team)
1. Manager of Incubation Center
2. Manager of Corporate Relations
3. Director of Financial Aid (member of Executive Team)
F. Director of Admissions (member of Executive Team)
Sales Reps
G. Registrar
H. Administrative staff
XIV. Process Applications
XV. Establish and Track Performance Metrics
XVI. Annually Review and Adjust the Plan

This "How To" guide comes in the form of a three ring binder that will be your companion throughout your design and implementation project. It contains a detailed description of each design and implementation step including numerous forms, tables and templates for your customization and use. A diskette containing all of these documents in native form for customization and printing is included in the Guide.

The Guide can be purchased directly from EII Publications. Contact me at rkw@eiicorp.com to order your copy and set up your annual subscription. Updates occur irregularly during the year. The first year updates are included in the purchase price. Thereafter you will be billed the annual subscription fee on your anniversary dates. This will continue automatically. You may cancel at any time. The updates will contain the experiences of adopting institutions so that you can benefit from the work of others. As new information becomes available it will be integrated into the Guide. At some point in time when a critical mass of adopting institutions is on board a Community of Practice will be formed with annual national and quarterly regional conferences offered.

RECOMMENDATION FOR A SUCCESSFUL IMPLEMENTATION

Every organization is different and offers the services that best meet their market needs.

The project you are about to undertake is being proposed to address your specific business situation. It has three phases. In the first phase you will design your own version of the WBDC Model that I have described in this book. Starting with my model as a template your model will evolve in a facilitated workshop format. You are going to design a customized version of my WBDC Model to take advantage of a heretofore untapped business opportunity. In the second phase you will develop a plan to implement your WBDC Model and in the third phase you will execute your implementation plan. None of this can happen until you have assessed your market situation.

I would like to spend 2 days with your core team and answer the critical questions. We will define your market needs and your expectations for serving that market. That needs analysis will be the input I need to craft a 3 day workshop for your team. The process described in this chapter is a robust process that works best when tailored to the institution's environment and needs and that is what I would produce using the information collected during the 2 day needs analysis. Creating a customized workshop from the information I gather in the needs analysis will maximize the use of our workshop time. Every implementation will be unique to your capabilities and your market situation. With that defined the deliverable from the 3 day workshop will be a detailed business and phased implementation plan. As workshop facilitator I will coach your team through the decisions they will have to make. The decision to go forward with the project can then be made by your team in confidence. You will know where you are, where you want to go and how you will get there. The next step is implementing the plan and I will be available to help with that if needed. My objective from the workshop is to give your team all of the tools it needs to execute a successful implementation. I'm not interested in creating a lifetime dependency for myself. I want you to become self-sufficient as soon as possible.

Alternatively once you have studied the Guide you and your team may feel confident that you can design and develop an implementation plan and proceed on your own. That's fine. Know that me and my team stands ready to offer whatever help and advice we can as you embark on your great adventure.

At the outset there should be a significant community awareness program utilizing all available media. This will include a number of presentations at Chamber of Commerce meetings, Kiwanis Clubs, professional society meetings, business clubs and other appropriate venues throughout the market area. The bottom line in this program is to garner financial and in kind support from the business community and private citizens. Creating visibility is a key to the successful implementation of the program.

Putting It All Together

As far as I know the WBDC Model that I have described in this monograph is unique. I envision it as a dynamic living program. In defining its contents and delivery process we, as educators and trainers, will be challenged to constantly re-invent ourselves and are limited only by our own creativity. Because the WBDC Model is based on a team-centric and project-based learning model it will automatically be aligned to the needs of business and produce graduates who have demonstrated through actual WBDC-based experiences that they can fill those needs. Having had this experience as part of their education and training is a powerful credential and should serve the worker as they enter the world of work.

...we, as educators and trainers, will be challenged to constantly re-invent ourselves and are limited only by our own creativity.

But the WBDC Model goes even further. It is designed to support the worker over their entire career. Things will change and technologies once thought to be necessary will be replaced by even more powerful technologies, new opportunities will arise and the cycle will repeat itself over and over again. Career and professional development is a lifelong journey. The WBDC Model will also adapt and be there for lifelong support of the worker. I have two projects underway to implement this WBDC Model. One is in New England and one is in the Caribbean.

If I have your attention and you would like to pursue this WBDC Model further, please contact me at ***rkw@eiicorp.com***. I would love to share my dream with you.

CHAPTER 8

Your Next Steps

I feel like we are at the threshold of a great adventure and I want to step into that future with you and your team. Our economy and society are crying out for a solution to the problems of the worker and the entrepreneur and there doesn't seem to be anything meaningful on the horizon. Are our leaders doing anything that the congress will approve? Maybe the strategy is to delay action, wait for something good to happen, and then claim the cure. I wonder.

The economy is a mess and no one would deny that. Our best minds are only guessing at cures. I have always believed that a hands-off approach by the government might be the best. Unfortunately it is too late for that approach. Capitalism and the markets are resilient and will recover of their own volition. It has happened before and it will happen again. The worker is challenged to prepare themselves for a career that has a future. But what is that career and what is needed to prepare for it? The entrepreneur is equally challenged. They need to find businesses and ways to implement those businesses so they won't be victims to technology and global competitors. Are there any sustainable businesses they can pursue? But what kind of businesses might they be and how can they proceed?

Neither the worker nor the entrepreneur can sit back and wait for a bailout. They have to take charge in difficult times and make things happen!

Society is a mess as far as the worker is concerned. Corporate loyalty to their workers is gone and there is no sign that it will ever come back. Workers are on their own and they are not prepared for the challenges they are facing. Jobs have disappeared or are moved off-shore. Technology has enslaved many people and they don't know how to counter. The worker owns their career and their employer owns their job. But the worker has to take charge and make things happen.

The worker owns their career and their employer owns their job. But the worker has to take charge and make things happen.

The higher education community has heard the clarion call for help and is socially and morally bound to step into the breach with programs to support the worker, the entrepreneur and the business owner. The time is right and we have an opportunity to make a lasting difference. But it will take a disruptive innovation like the WBDC Model to make a lasting difference.

If my presentation of this WBDC Model makes sense and you would like to implement it, please let me hear from you. I am most anxious to work with you to turn the WDP Model into reality in your community. At the least I can offer you my own experiences and advice. This will not be an easy journey for many because it goes to the very roots of how we train and educate our workers and support their career and professional development journey. Change in the educational delivery systems will not be easy but I offer my WBDC Model as the only way that really makes sense.

This is my opening gambit at creating sustainable solutions and it will not be the last attempt but it is a start. Like any disruptive innovation it will require a number of revisions as it matures and makes the contribution it was originally purposed to deliver.

WHAT SHOULD YOU DO NOW?

When business is in a downturn management gurus suggest that it is a good time to do strategic planning in preparation for the eventual upturn in business. The same logic applies to the markets that the WBDC serves. The questions posed earlier and reproduced below serve as a good outline of your strategic plan.

- *Where are you?* What you are doing now for the worker and the entrepreneur is the beginning of your strategic plan. A description of your competitors, the market and how well the market is being served should also amplify where you are.
- *Where do you want to go?* The answer begins with a statement of what could be done to better serve your market. You might want to meet all the needs of the market or perhaps only a subset of those needs. A good capabilities statement might facilitate that answer.
- *How will you get there?* This is your strategic plan. You might proceed on your own based on the materials presented in this book or you might engage us to work with you to plan and implement your own WBDC Model. The Guide outlined below is a companion to this book and provides a complete "How To" document for your use.

It outlines the project as we would undertake it on your behalf.

My team is prepared to join your team and make the WBDC Model a reality in your market. I can be reached at rkw@eiicorp.com.

Appendix A

Selected Funding Sources

There are over 100 public, private and corporate agencies and organizations that have supported projects related to the WBDC concept. This section contains fundamental information on a few of these. There are many more. This list will change continuously and every attempt will be made to keep this list current. This list is only a sample of the kind of information you can expect to find annotated in the Guide. The Guide also provides more details on all the other funding sources for WBDC-related programs. Any information that I have found that is related to applications will be included too. The listing is presented in random order.

Table A.1. National Science Foundation

Item	Information
Name	National Science Foundation
URL	www.nsf.gov
Mission	The goals of the Partnerships for Innovation Program are to: 1) stimulate the transformation of knowledge created by the research and education enterprise into innovations that create new wealth; build strong local, regional and national economies; and improve the national well-being; 2) broaden the participation of all types of academic institutions and all citizens in activities to meet the diverse workforce needs of the national innovation enterprise; and 3) catalyze or enhance enabling infrastructure that is necessary to foster and sustain innovation in the long-term. To develop a set of ideas for pursuing these goals, this competition will support 12-15 promising partnerships among academe, the private sector, and state/ local/ federal government that will explore new approaches to support and sustain innovation.
Type of Support Offered	
Size and Duration of Funding	
Application Dates	
Comments	

Table A.2. Ford Foundation

Item	Information
Name	Ford Foundation
URL	www.fordfound.org
Mission	The Ford Foundation supports visionary leaders and organizations working on the frontlines of social change worldwide. Our goals for more than half a century have been to: * Strengthen democratic values * Reduce poverty and injustice * Promote international cooperation * Advance human achievement A fundamental challenge facing every society is to create political, economic and social systems that promote peace, human welfare and the sustainability of the environment on which life depends. We believe that the best way to meet this challenge is to encourage initiatives by those living and working closest to where problems are located; to promote collaboration among the nonprofit, government and business sectors; and to ensure participation by men and women from diverse communities and all levels of society. In our experience, such activities help build common understanding, enhance excellence, enable people to improve their lives and reinforce their commitment to society. We work mainly by making grants or loans that build knowledge and strengthen organizations and networks. Since our financial resources are modest compared with societal needs, we focus on key problem areas and program strategies. Created with gifts and bequests by Edsel and Henry Ford, the foundation is an independent, nonprofit, nongovernmental organization, with its own board, and is entirely separate from the Ford Motor Company. The trustees of the foundation set policy and delegate authority to the president and senior staff for the foundation's grant making and operations. Program officers in the United States, Asia, Africa and Latin America explore opportunities to pursue the foundation's goals, formulate strategies and recommend proposals for funding.

Type of Support offered	Start-up support for the National Fund for Workforce Solutions to strengthen and expand high-impact workforce development initiatives through technical assistance, grant making, research & advocacy.
	Each year the Ford Foundation receives about 44,000 proposals and makes some 2,000 grants. Types of support include grants, recoverable grants, loans and loan guarantees. Requests range from a few thousand to millions of dollars and are accepted in categories such as project planning and support; general support; and endowments. Grant applications are reviewed at our New York headquarters and in our regional offices. Please check the locations of our regional offices to determine if we operate in your country of interest.
	Grants
	A grant is a commitment by the foundation to make payments to an organization or an individual over a set period of time to further one of the interest areas in which we work. The foundation gives the grantee autonomy over management of the funds, but all grantees must sign a letter agreeing to abide by the terms and conditions of the grant. Grants administrators ensure that the grant-making process—from preparation of the grant recommendation to the closing of the file—conforms to the foundation's procedures and standards. Here are the types of grants we make:
	* General/core support
	* Project
	* Planning
	* Competition
	* Matching
	* Recoverable
	* Individual
	* Endowment
	* Foundation-administered project
Size and Duration of Funding	Multiple year but funding levels not specified
Application Dates	Continuous

Comments	We make grants to develop new ideas and strengthen organizations that reduce poverty and injustice and promote democratic values, international cooperation and human achievement. To achieve these goals, we take varied approaches to our work, including supporting emerging leaders; working with social justice movements and networks; sponsoring research and dialogue; creating new organizations; and supporting innovations that improve lives. These methods of problem-solving reflect our values and the diverse ways in which we support grantees. They also describe a model of philanthropy that the foundation has pursued for more than 70 years: to be a long-term and flexible partner for innovative leaders of thought and action. Lasting change in difficult areas, such as the reduction of poverty, protection of human rights and establishment of democratic governance after a dictatorship, requires decades of effort. It involves sustained work with successive generations of innovators, thinkers and activists as they pursue transformational and ambitious goals. Our mission is broad, and we carefully target how our grants can be used most effectively. Once the foundation decides to work in a substantive or geographic area, our program staff consults with practitioners, researchers, policymakers and others to identify initiatives that might contribute to progress. We explore specific work grantees might undertake, benchmarks for change and costs. A program officer conducts this analysis, and then presents the ideas in a memorandum reviewed by peers, a supervisor and at least two foundation officers. When it is approved, the program officer begins to make grants within the broad parameters of the approved memorandum and a two-year budget allocation. Staff members regularly provide reports to the board about grants made and ongoing lines of work. Through the foundation's Web site and publications, the public learns of program initiatives and emerging areas of interest.

Table A.3. Microsoft Corporation

Item	Information
Name	Microscoft Corporation
URL	www.microsoft.com
Mission	Reach over 20 million people globally per year through the Community Technology Skills Program.

Reach 250 million students and teachers across 115 countries through our $500 million investment in the Partners in Learning program by 2013.

Provide job skills training to 2 million people by 2012 through Elevate America, and to 500,000 people by the end of our 2010 fiscal year.

Increase the number of students globally that benefit from the Students to Business program from 100,000 in 2009 to more than one million by 2012.

Innovation is the cornerstone of the success, of Microsoft, and it is central to our value in society. Our growth depends on developing new technologies, entering new geographic areas and product markets, and encouraging the use of our products and services. We attract top talent to our research facilities in countries and regions around the world, including Canada, China, Denmark, England, Ireland, India, Israel, and the United States. Our broad reach allows us to work locally addressing challenges and developing markets. We are investing $9.5 billion in research and development this year alone. |
Type of Support Offered	
Size and Duration of Funding	
Application Dates	
Comments	

Table A.4. Intel Corporation

Item	Information
Name	Intel Corporation
URL	
Mission	
Type of Support Offered	Intel will provide grants, donations, and philanthropic support to additional programs that improve the quality of life in its site communities. Each request will be evaluated on the basis of the services offered and the program's impact on the community, its focus on diversity and multiculturalism, its impact on the youth of our community, the cost-effectiveness of the program and its ability to be effectively measured and replicated, and the potential for Intel employee involvement.
Size and Duration of Funding	
Application Dates	
Comments	

Table A.5. Staples Foundation for Learning

Item	Information
Name	Staples Foundation for Learning
URL	www.staplesfoundation.org/foundationhome2.html
Mission	To teach, train and inspire people from all walks of life by providing educational and growth opportunities
Eligibility	The organization must: * have a nonprofit tax-exempt classification under 501(c)(3) of the Internal Revenue Code. * align with Staples Foundation for Learning's mission and give focus on job skills and education.
Type of Support Offered	Of the proposed amount Staples will allocate 80-100% for program expenses and 0-20% for operational expenses
Size and Duration of Funding	Not Specified
Application Dates	Not specified. Awards are given three times per year.
Comments	Staples Foundation for Learning does not accept unsolicited proposals. Staples Foundation for Learning has developed lasting relationships with Ashoka, Initiative for a Competitive Inner City, Earth Force and Hispanic Heritage Foundation, these non-profit organizations focused on global youth entrepreneurship, inner-city businesses, the environment and Hispanic youth respectively. In addition, Staples Foundation for Learning provides grants to hundreds of local grassroots organizations.

Table A.6. Lumina Foundation

Item	Information
Name	Lumina Foundation
URL	www.luminafoundation.org
Mission	The mission of Lumina Foundation for Education is to expand access to postsecondary education in the United States. The Foundation seeks to identify and promote practices leading to improvement in the rates of entry and success in education beyond high school, particularly for students of low income or other underrepresented background. It likewise seeks improvement in opportunities for adult learners. The Foundation carries out the mission through funding and conducting research; communicating ideas through reports, conferences and other means; and making grants to educational institutions and other nonprofits for innovative programs. It also devotes limited resources to contributing appropriately in support of selected community and other charitable organizations.
Type of Support Offered	Intel will provide grants, donations, and philanthropic support to additional programs that improve the quality of life in its site communities. Each request will be evaluated on the basis of the services offered and the program's impact on the community, its focus on diversity and multiculturalism, its impact on the youth of our community, the cost-effectiveness of the program and its ability to be effectively measured and replicated, and the potential for Intel employee involvement.
Size and Duration of Funding	
Application Dates	
Comments	

Appendix B

Curriculum & Core Courses ━━━━━━━━━━━━━━━━━━━━━

These 11 core courses total 17 credit hours and could form an Entrepreneurship Certificate Program or a minor within a BA or BS program in any discipline. By adding a few electives a BA or BS in Entrepreneurship can be defined. Other formal courses will be drawn from existing courses offered by the participating institutions. In collaboration with the participating institutions students will file and gain approval of degree or certificate programs according to the program requirements of their home institution. Due to the adaptive nature of the programs there will be a heavy reliance on project-based and problem-based courses. The content of these courses will be adaptive but still maintain compliance to their home institution's program requirements. Because of the adaptive nature of the WDP Curriculum these will be defined at the appropriate time in each student's program.

I don't want you to get the impression that the WDP Curriculum is a free for all curriculum. It is adaptive because it uses project and problem-based learning models. Courses derive their content and structure from the projects and problems contributed by businesses and others but there is an underlying learning requirement defined by their program requirements. So there is a standard that defines what disciplines, concepts, principles and theories must be included in the certificate or degree program but not how that is included in the program.

All 11 courses listed below are designed for a team-based, project-driven curriculum. Student teams bring BIC projects with them which they will execute as part of meeting their learning objectives that have been documented in their SLC. The WBDC Curriculum consists of the following core and entrepreneurship courses:

General Core Courses
- Career & Professional Development Planning Workshop
- Introduction to Entrepreneurship
- Fundamentals of Effective Project Management
- Fundamentals of Business Analysis
- Creative Problem Solving Methods
- Written & Verbal Communications Skills
- Effective Team Formation & Building

Entrepreneurship Courses
- Fundamentals of Creativity and Innovation
- Marketing Research
- How to Write a Business Plan
- Financing the New Venture

The General Core Courses are introductory to all programs in the WDP. All six courses are open to the public. In consultation with an advisor the actual required core courses that will be required of the student as a condition of acceptance into the WDP will be identified. The four Entrepreneurship Courses when added to the General Core Courses form a concentration in Entrepreneurship that can be accommodated into certificate, associate, and bachelor degree programs.